Fundamentals of
Occupational
SAFETY and HEALTH

Dr. James P. Kohn, EdD, CSP, CIH, CPE
Associate Professor

Dr. Mark A. Friend, EdD, CSP
Associate Professor

Dr. Celeste A. Winterberger, PhD
Assistant Professor

Industrial Technology Department
East Carolina University
Greenville, North Carolina

 Government Institutes, Inc. • Rockville, Maryland

Government Institutes, Inc., 4 Research Place, Suite 200, Rockville, Maryland 20850

ISBN: 0-86587-539-1

Printed in the United States of America

Summary of Contents

Table of Contents

Chapter 6. Introduction to Industrial Hygiene 99

List of Figures and Tables

Foreword

Drs. Kohn, Friend, and Winterberger have written a book that is principally directed to the needs of a student in a safety technology and management course. It can also serve as a primer for the person who has no previous safety education but has been assigned the responsibility for safety. Additionally, the safety professional who needs information and answers on a particular subject will find this book to be a valuable reference.

Their book is a well-balanced presentation of the fundamental safety technology and management needs of the safety professional as they continue to evolve. There are several driving forces affecting the changing content of the practice of safety. Among them are the

1. Combination in many organizations of the staff resource responsibilities for environmental, safety, and health matters into one entity;
2. Emergence of ergonomics to become a far more significant aspect of the safety practitioner's work;
3. Awareness that knowledge of systems safety is becoming a requisite for effective safety practice; and
4. Greater attention being given by management to workers' compensation costs.

Treatment of safety, industrial hygiene, and environmental fundamentals in this book is, therefore, necessary and appropriate, as the knowledge requirements for the safety position have been broadened in many organizations.

Ergonomics continues to grow in importance within the practice of safety because of the mounting evidence that ties good ergonomic design for safety to improved productivity and cost efficiency. Ergonomics is fundamentally design based, as are the needs for productivity and lower unit costs. It is difficult to visualize the occupational safety practitioner being successful in the future without a good grounding in applied ergonomics.

Slowly, awareness grows of how significant the concepts of system safety will be in the general practice of safety. In the first chapter, the authors discuss the *Scope and Functions of the Professional Safety Position,* a document issued by the American Society of Safety Engineers. The document cited is a recent revision. It contains some subtle but important changes that indicate how the practice of safety is evolving.

This publication sets forth four "major areas relating to the protection of people, property, and the environment." The first two are: "Anticipate, identify, and evaluate hazardous conditions and practices;" and "develop hazard control designs, methods, procedures, and programs."

A previous edition of the scope and functions document did not include the words *anticipate* or *designs* in these two subject areas. To be able to anticipate hazards, one must be involved in the design process. Anticipating, identifying, and analyzing hazards in the design process so that they can be eliminated or controlled is what system safety is all about. The system safety concepts set forth in this book apply to every technology and management aspect of the practice of safety.

All of the foregoing are affected by legislation, which is dealt with in the early chapters. It would be a rare safety position that did not require a thorough knowledge of the applicable legislation, including the requirements for workers' compensation to which far greater attention is now being given. These early chapters cover important and fundamental needs.

Particular attention is drawn to the chapter on accident causation theory and accident prevention. Because incident investigation is an important role for most safety practitioners, a thorough understanding of causation theories is a necessity. Incident investigation is an aspect of the practice of safety which can be improved.

In addition to having knowledge of environmental, safety, and health technology, a safety practitioner must know how to manage a safety function and must have a realization of how things get done with and through people. These subjects are vital and well covered herein.

Fred A. Manuele, CSP, PE
President
Hazards, Limited

Preface

Frustration caused us to write this book. We had spent the better part of our careers as safety educators looking, in vain, for a basic text in occupational safety and health which accomplished a few simple goals:

- The book should not assume the reader has experience or prior exposure to occupational safety and health, industrial hygiene, or occupational medicine.

- It should be written in plain English and it should explain the basic principles behind eliminating workplace hazards and protecting workers from potential hazards encountered as a result of their professions.

- And topics too advanced for someone trying to understand the basics of safety and health should not be included.

We knew what we wanted to see covered in this book, but were our ideas shared by others? We decided to ask both safety and health professionals and our academic colleagues what they wanted to see in such a book, how they would prefer it to be organized, what topics should be stressed, and which should be handled in summary fashion. Their responses were virtually identical to the issues and topics we ourselves identified. The result of those surveys is that this is a book that many safety professionals wish they'd had when they first entered the profession and that many safety veterans will still be able to learn a thing or two from.

As the title implies, *Fundamentals of Occupational Safety and Health* covers the basics safety and health professionals need to understand before they can accept responsibility for reducing hazards and protecting the lives entrusted to them. Unlike books which are available which only explore the engineering aspects of OS&H in depth, our goal is to balance the management of safety with the relevant science and the practical aspects of complying with regulations.

It can be used by students and faculty at traditional safety programs, as well as business schools and industrial technology programs where an introduction to safety is required. It will be especially valuable, in this age of downsizing and corporate and governmental reengineering, to the countless environmental professionals struggling to integrate employee safety and health concerns into their existing responsibilities.

As the role of government in insuring employees' right to a workplace free of hazards is debated, it will be more important than ever for safety and health professionals to reflect on and return to the basics. We hope that this book will make a contribution to the improvement of our profession and bring an end to many current and future safety professionals' frustrations.

James P. Kohn, EdD, CSP, CIH, CPE

Mark A. Friend, EdD, CSP

Celeste A. Winterberger, PhD

About the Authors

James P. Kohn is associate professor of Industrial Technology at East Carolina University and vice president for professional development of the American society of Safety Engineers and President of Occupational Safe Service Corporation. He was previously director of the Occupational Safety Management program at Indiana State University, holds a doctorate in education, and is a Certified Safety Professional, Certified Industrial Hygienist, and Certified Professional Ergonomist. Dr. Kohn has written three books on safety and industrial hygiene topics. He is a member of the American Society of Safety Engineers, American Industrial Hygiene Association, and the American Conference of Governmental Industrial Hygienists.

Mark A. Friend is director of East Carolina University's Occupational Safety and Health Consortium and of the Center for Applied Technology. He is also an associate professor of Industrial Technology at East Carolina University. Dr. Friend is the former chairman of the Occupational Safety and Health department at Murray State University, is past president of the National Occupational Safety and Health Educators' Association, and is a member of the American Society of Safety Engineers.

Celeste A. Winterberger, is an assistant professor in the Department of Industrial Technology at East Carolina University. She received a B.S. in Zoology from Iowa State University, a B.S. in Industrial Technology from Northern Iowa, a M.S. in Industrial Technology from Purdue University, and a Ph.D. in Industrial Education from Texas A&M University. She was previously an assistant professor in the Department of Industrial Technology at Illinois State University.

Acknowledgments

The authors wish to thank Glory Mizzelle, Darla Hinnant, Doug Gaylord, David McDaniel, and Bill Benfield for their contribution to the development and editing of the manuscript. Thanks also go to the following individuals for their gracious assistance in providing information, guidance, and text for specific chapters as listed:

Chapter Two:	Tom Wagner, Craig MacMurray, Sam Christy, Kelly Nelson, and Wes Heinold
Chapter Three:	Tom Wagner, Susan Wilson, and Craig MacMurray
Chapter Four:	George Nichols, Jr.
Chapter Five:	Eddie Allen, Eddie Anderson, Barbara Dail, Ralph Dodge, Doug Gaylord, Eddie Johnson, Barry Maxwell, Amy Tomlin, and Celeste Winterberger
Chapter Seven:	Eddie Anderson, Barbara Dail, Ralph Dodge, Doug Gaylord, Eddie Johnson, Katherine Kohn (ergonomic illustrations), Barry Maxwell, Amy Tomlin, and Celeste Winterberger
Chapter Eight:	Amber Perry
Chapter Twelve:	Brett Carruthers (who co-authored this chapter with Mark Friend)
Chapter Thirteen:	Tom Shadoan, Antoine Slaughter, Julie Mitchell, Jeff Porter, James Frey, and Joseph Johnson
Chapter Fourteen:	Robert Getchell, Keith McCullough, Mike Van Derven, and Neil Brown
Appendix A:	Eddie Anderson, Mike Baker, Barbara Dail, Craig Fulcher, Doug Gaylord, Barry Maxwell, David McDaniel, and Ron Skinner
Appendix B:	Mike Baker, Craig Fulcher, Doug Gaylord, Darla Hinnant, David McDaniel, Glory Mizzelle, Ron Skinner, Ward Taylor, Chip Tillett, and Jason Whichard.

In addition, the authors want to recognize Carrie Kohn, M.A., and Sandy Friend, M.S., as well as Mr. and Mrs. R.E. Winterberger for their professional encouragement.

Chapter 1

Introduction to Occupational Safety and Health

CHAPTER OBJECTIVES

After completing this chapter, you will be able to

- Explain the importance of occupational safety and health.
- Identify key historical figures that have contributed to the profession.
- Define terminology used in occupational safety and health.
- List job titles of individuals that perform occupational safety and health activities.
- Identify roles and responsibilities of safety and health professionals.

CASE STUDY

As a twenty-two-year-old construction worker with eleven months experience on the job, Bob had finally made it. Since graduating from high school, Bob had tried a lot of things, but they just never seemed right for him. He had gone to a community college for one year, but he ended up dropping out of school. There was too much theory which didn't relate to how he saw the world. Bob then tried a number of jobs, but minimum-wage salaries forced him to live at home with his parents. They were good people, but he was ready to move on in his life. With this new job, everything was turning out great. Bob was bringing home a good paycheck. He had just moved into a new apartment, which he shared with his high school buddy Tim, and he was going the next day to sign the papers for a brand new pickup truck.

Bob never made it to the dealership to sign those papers. Maybe he was distracted thinking about that "killer" pickup that he was about to purchase. Perhaps he never realized how dangerous it actually was to be working up on that scaffolding. After all, it was only twenty feet off the ground and it looked safe. Sure it was wet from the rains the night before, but he had worked on wet scaffolding before and nothing ever happened to make him concerned about working at those heights. For whatever reason, Bob's world did change when he fell those few feet to the ground. Bob became a statistic, another number added to the fall statistics for construction work in the United States. His fall put him in a wheel chair, paralyzed from the waist down.

OCCUPATIONAL SAFETY AND HEALTH

What is occupational safety and health? Occupational safety and health is the discipline concerned with preserving and protecting human and facility resources in the workplace. Practitioners in this field are concerned with preventing needless deaths and injuries of workers like Bob. It is a profession, however, that involves much more than first-aid activities. It is a profession that is far-reaching in both scope and practice. For some individuals, occupational safety and health is concerned with preventing people from being injured or becoming ill due to hazards found in the places where they work. Concern for Bob and ensuring that he was trained in hazard recognition, as well as the correct methods for performing his job, would be an example of this.

For others, occupational safety and health is the area where professionals attempt to prevent catastrophic losses. They prevent conditions which can result in disasters that affect the workplace. Safety professionals are attempting to prevent explosions or fires that can destroy entire buildings when they conduct fire inspections. Some professionals consider occupational safety and health as the management function in an organization concerned with improving its quality and efficiency. Safety professionals who look at the way products are moved from place to place are concerned with both reducing lifting injuries exposure, while at the same time minimizing product movement. This is achieved through the elimination of property and facility damage, waste and costs that take away from the organization's ability to operate profitably.

In practice, occupational safety and health includes all of these moral and economic issues. In fact, there is also a legal reason for companies to promote occupational safety and health. The United States government, as well as most governments around the world, require that people be protected from those hazards in the workplace that can result in injuries, illness, or death. Under the Occupational Safety and Health Act of 1970, commonly known as the OSHAct, every employer in the United States is required to provide a safe and healthful workplace.

It is clear that companies are required by law to ensure that their employees will not be placed in dangerous environments. For many employers, however, the responsibility to protect human life is not as important as other goals. Some individuals are focused only on productivity and profits. They view illnesses and injuries as a part of the job. In reality, the opposite case can be made. The amount of production required to cover costs associated with accidents in the workplace can be substantial. In a company that manufactures windows, the president stated, "We must make and sell 9,000 windows to pay for just one... 'wrist'... surgery including lost work days, workers' compensation, and all related expenses" (Strakal, 1994). Refer to the discussion on cost-benefit analysis for an in-depth examination of this issue.

IMPORTANCE OF OCCUPATIONAL SAFETY AND HEALTH

Economically, morally, and legally, occupational safety and health has become an important issue in today's society. Companies are attempting to remain profitable in an ever-more-competitive, global economy. For these companies, addressing safety, health, and environmental issues may mean more than good business practice. For many companies strong safety, health, and environmental programs may actually mean survival. Occupational safety and health statistics point to the extent of this problem.

According to the National Safety Council's 1994 publication entitled *Accident Facts*, over $113.9 billion was lost as a result of occupational injuries. The staggering numbers of people involved in these accidents are a real concern. Statistics in occupational safety and health revealed that in 1993, 3,200,000 workers were injured and over 9,000 workers were killed as a result of workplace accidents. These are numbers. Behind these cold, faceless numbers are real people: mothers, fathers, sisters, brothers, spouses, sons, or daughters. They are the people like Bob whose lives may never be the same again.

How often are people in the United States personally affected by occupational safety and health incidents? Approximately 3.5 of every 1000 people will die as a result of occupational fatalities including work related vehicle accidents, workplace violence, or flammable gas explosions. The average cost to society for each of these fatalities will be $780,000 per victim. One out of three Americans will suffer a workplace injury that results in medical expenses averaging $940 as a result of machinery accidents, falls, or being struck by falling objects. Every ten minutes approximately four people will have serious injuries on the job.

Have you known friends or family members seriously injured at their jobs? Not the nonserious cuts or bruises that people get performing daily activities, but the serious injuries and illnesses requiring medical attention from trained professionals. Many of you may have been personally affected by the loss of loved ones killed while performing their jobs. Statistically, during the next four years, every single working American will be involved in a serious occupational injury. What is especially shocking is that these numbers do not address off-the-job travel or home safety issues.

The Occupational Safety and Health Administration (OSHA), the federal agency responsible for workplace safety and health, is stepping up its activities in response to these statistics and to the safety concerns being expressed by Americans. OSHA is pursuing civil action and seeking criminal prosecution of business owners and managers who willfully neglect the safety and health of their employees. These actions are being taken because Americans are serious about their personal health and well-being.

Often, these actions are typically in response to disasters and catastrophes. This issue will be further examined in the next chapter covering occupational safety and health legislation.

HISTORICAL EXAMINATION OF OCCUPATIONAL SAFETY AND HEALTH

Ancient Greek and Roman Physicians

Some people in our society believe that concern for occupational safety and health is a relatively recent issue. In fact, it has a relatively long and important history. Why is the study of occupational safety and health history important? Many of the health and safety hazards of concern today were first observed over two thousand years ago. The earliest account of this interest is associated with the Code of Hammurabi which dates back to approximately 2100 BC. The Code of Hammurabi was primarily concerned with personal injury and losses. A detailed examination of this code will be discussed in a later chapter.

Many of the earliest reported observations of workplace safety and health issues were made by ancient physicians. Greek and Roman physicians, practicing between 400 BC and 300 AD, expressed concern for the health of individuals exposed to the metals commonly used during this period. Hippocrates, the Father of Medicine, was the ancient Greek physician who is perhaps best known for the "Hippocratic Oath" taken by graduating physicians. Practicing around 370 BC, he was one of the first physicians that recognized and reported lead poisoning in the mining industry. People conquered by these ancient societies were often required to provide "slave labor" for the mining of mercury, lead, and other valuable metals of the times. Hippocrates told his students to consider environmental and work factors when making a diagnosis, especially for the refiners and extractors of metals.

Pliny, the Elder, a Roman physician and scientist, was another contributor to the ancient knowledge associated with safety and health. Living between 23 and 79 AD, Pliny gave his life to science when he died studying the eruption at Mount Vesuvius. It is believed that he died when he was exposed to sulfur fumes expelled during the eruption that buried Pompeii. This was the natural disaster that killed thousands of Roman citizens. Pliny recorded observations concerned with the hazards involved in handling lead, silver, and cinnabar (red mercuric sulfide). In his encyclopedia, he referred to the use of animal bladders worn as respirators which slaves used to avoid inhaling dust in mining and refining activities.

Another individual who recognized occupational safety and health issues was Galen. He was a Roman physician who lived during the second century. He wrote about occupational diseases and the dangers of acid mists to copper miners. Galen was most concerned with the mining, tanning, and chemical occupations, noting several diseases contracted by individuals working in those professions.

European Renaissance and the Industrial Revolution

Little information is available on European injury, illness, and property damage prevention activities before the end of the fifteenth century. Reports of medieval scribes suffering lead poisoning while performing the common practice of tipping their quills with their tongues between dips into metallic ink solutions was repeatedly noted. Unfortunately, little else was recorded regarding safety and health during that period in history. It was not until the European Renaissance that physicians and chemists once again began noticing the relationship between occupational activities and worker health and safety.

Ulrich Ellenborg, for example, recognized, identified, and reported "on the poisonous and noxious vapors and fumes of metals." In 1437 he recognized that the vapors of some metals, including lead and mercury, were dangerous and described the symptoms of industrial poisoning from these sources. He also became aware of asbestos and lung diseases among miners.

Another important contributor to the understanding of occupational safety and health during this period was Paracelsus. A Swiss chemist, physician, and professor of physics and surgery, whose birth name was Bonbatus Von Hohemheim, Paracelsus lived between 1493 and 1591. His publication, *Von der Bergsucht and Anderen Bergkrankheiten*, studied various miners' diseases in the mines surrounding Tyrol. He reported symptoms including disturbances of the lungs, stomach, and intestines, resulting from mining and refining gold, silver, alum, sulfur, lead, copper, zinc, iron, and mercury. His work forms the basis of the modern study and science of toxicology, "the study of poisons." In his publication, Paracelsus said, "All things are poisonous and none inherently poisonous. Only a dose determines severity." This concept of concentration and duration of exposure of contaminants is the basis for the Permissible Exposure Limits (PELs) (established by OSHA) and Threshold Limit Values (TLVs) (recommended by the Conference of Governmental Industrial Hygienists or ACGIH) for hazards substances.

While there are numerous other individuals during this period that could be discussed, the last individual that will be mentioned is Bernardo Ramazzini. Ramazzini was an Italian physician who published a book around 1700 titled *De morbis artificum diatriaba* or *The Diseases of Workers*. This could be considered the first treatise on occupational disease. Considered by some to be the "Father of Occupational Medicine," while others bestow upon him the title of the "Father of Industrial Hygiene," he recommended that physicians ask their patients, "What trade are you in?" He urged his students to learn the nature of occupational diseases in shops, mills, mines, or wherever men toil.

During the Industrial Revolution, the period between 1760 and 1840, history witnessed dramatic advances in technology. With this advancing technology came an increase in safety and health hazards. The innovations of mechanical textile machinery, foundry furnaces, steam engines, and numerous other inventions created a new workplace environment. Workers were exposed to workplace hazards that humans were never exposed to, nor aware of, before. Factories and other workplaces were a maze of moving belts, pulleys, and gears. Human senses were assaulted with fumes, toxic vapors, noise, and heat. The health and safety problem was compounded by the introduction of increasing numbers of women and children into the workforce. Long twelve-hour workdays, unsanitary conditions, and demanding physical labor increased the likelihood of injury and illness for this new workforce.

During this period in history (circa 1775), Dr. Percival Pott identified the first form of occupational cancer. He observed scrotal cancer in chimney sweeps and determined its relationship to soot and coal tar exposure. This initiated numerous regulations called the "Chimney Sweep Acts" that were promulgated between 1788 and 1875. During this period, several industrialists also became concerned with the welfare of their workers. Sir Robert Peel, for example, was a mill owner who made the English Parliament aware of the deplorable working conditions that often existed in many of the mills. Orphan labor was frequently used to perform demanding tasks in less than sanitary conditions. He aided in the study of these deplorable conditions, revealing that the mean life expectancy of the working class, under these terrible conditions, was only 22 years, while the mean age of the wealthier classes was 44 years.

Another key individual concerned with occupational safety and health during this period in history was Charles Thachrah. He studied the effects of arts, trades, life habits, civic states, and professions upon health and longevity. In addition, Edwin Chadwick, a British lawyer and sanitarian, was another individual who described the deplorable conditions of factory workers.

During the 18th and 19th centuries, occupational safety and health became an important area of concern in the United States. Americans were facing the same problems as their coworkers across the Atlantic. Massachusetts was one of the first industrial areas in the United States. It was in mill towns, like Lowell, Massachusetts, where young women and girls were often injured and maimed in the moving gears and pulleys of the textile machinery. They worked long hours, often five in the morning until seven in the evening. Many of these girls were as young as six to ten years of age. They were required to work with their hands very close to the in-running gears of spinning machines. Their fingers were cut off or mangled with such frequency that machine guarding laws were eventually passed. Fatalities in the mining and steel industries were as common as the textile injuries.

This brief historical review was meant to serve several purposes. First, it is important to recognize that many health and safety issues have been around for a long time. Exposure to toxic metals like mercury and lead has been an occupational health problem almost as long as people have worked. The second purpose of this review was to paint a picture of the way our ancestors lived and toiled in their working environment. Technological advances usually bring with them new and unique hazards that are often overlooked by the untrained observer. It is easy to be unaware of a problem until it has been around long enough to affect a lot of people.

Just as in the past, we continue to witness dramatic changes in technology and workplace design. These "technological advancements," however, have not necessarily resulted in healthier or safer workplaces. In many respects, workers are exposed to as many workplace hazards today as their ancestors were in years past. For example, the factory machinery with its unguarded gears that mangled fingers and hands have been replaced by electronic office equipment that can cause wrist and arm injuries.

Occupational safety and health is more important today than at any other point in human history. The complexity of today's safety and health hazards mirror the complexities associated with modern workplace technology. Practitioners in this profession must develop the broad range of knowledge and skills necessary to ensure the protection of people and company resources. This knowledge base must include a well-grounded understanding of the terms and concepts used in the profession. In addition, occupational safety and health professionals must possess the skills required to effectively perform their roles and responsibilities.

TERMS AND CONCEPTS IN THE SAFETY PROFESSION

As previously mentioned, safety professionals are individuals responsible for people and company resources. They are individuals who have the knowledge and skills that are often acquired through formal education and/or experience in the safety field. Many safety professionals are certified after successfully completing the requirements for the designation Certified Safety Professional (CSP), Certified Industrial Hygienist (CIH), or other related occupational safety and health certifications.

Safety professionals attempt to achieve their loss prevention goals through the systematic use of principles taken from a variety of disciplines. Disciplines that play an important role in the safety profession include engineering, education, psychology, physiology, industrial hygiene, health physics, and management. Safety professionals are concerned with the elimination or control of hazards that may result in injury, illness, and property damage. They will often use techniques referred to as "loss prevention" and "loss control" to accomplish that goal.

Loss prevention describes a program designed to identify and correct potential accident problems before they result in financial loss or injury. *Loss control*, on the other hand, is a program designed to minimize incident-based financial losses. An example of the difference between loss prevention and loss control can be seen in the various activities associated with a fire protection and prevention program.

In a fire prevention program, employees can be trained to inspect their areas and remove combustible materials like oily rags or cardboard. These inspection activities would be an example of loss prevention. A fire protection program, on the other hand, might include employee training in the use of fire extinguishers. Employees would then possess the skills necessary to fight a fire were one to start on the shop floor. The fire might ignite, but, following training, employees are prepared to extinguish it and thus minimize the damage. Fire extinguisher training would be an example of loss control.

Loss prevention and loss control techniques are important to the safety professional who attempts to recognize, evaluate, and control hazards in the workplace. This is part of the process referred to as safety management. Safety management encompasses the responsibilities of planning, organizing, leading, and controlling activities necessary to achieve an organization's loss prevention and loss control goals. Continuing the fire example, the safety professional might wish to establish the safety management program to address this danger. Safety professionals would determine the problems that exist at their facility. They would then establish the details of the fire training program with goals and objectives of what is to be accomplished (planning). Next, they would determine the trainers and materials necessary to implement the program by establishing a schedule to ensure that all activities are accomplished (organizing). Safety professionals must then ensure that the required resources are available and that the people involved in this project coordinate their efforts when required (leading). Finally, the professional monitors and evaluates the progress of the project (controlling). A more detailed examination of the management of the safety function will be presented in chapter ten.

One of the most important terms used in the safety and health profession is *safety*. It is probably the most misinterpreted term by individuals outside of the safety profession. For the lay person, safety means not getting injured. This is not the way that professionals define the term. *Safety*, to the professional, denotes the likelihood or risk that a loss event will occur. It can be defined as "an acceptable or low probability of risk associated with conditions or activities with the potential to cause harm to people, equipment, or facilities."

There are several different types of occupational health and safety losses that safety professionals attempt to eliminate or control. Typical worker-related health and safety losses include: injuries, illnesses and fatalities.

Workplace losses can include: damaged equipment, damaged raw materials or finished products, damaged or destroyed facilities, downtime, service/production interruption, or loss of reputation.

When discussing losses, an understanding of the terms *risk* and *hazards* is of vital importance. *Risk* can be defined as the measure of the probability and severity of a loss event taking place. *Hazards* are the workplace conditions or worker actions that can result in injuries, illnesses, or other organizational losses. As revealed in these definitions, to determine occupational risks requires safety professionals to examine both the probability of an event occurring along with the potential severity of the end result.

The evaluation of risks in the workplace start with the identification of the types of hazards that exist at the facility. Establishing a process that ensures that hazards are identified is the primary goal of progressive organizations with good safety management programs. The organization then eliminates or reduces the risks associated with those hazards to the lowest achievable and reasonable level. It must be pointed out that nothing is risk free. Safety professionals identify the tasks and activities having the greatest inherent risk. They then attempt to systematically eliminate or reduce the level of risk as much as is feasibly possible given time, personnel, and budget restraints. A detailed discussion about risk assessment will be presented in Chapter 5, examining Accident Causation Theory and Accident Investigation.

The final term of importance to be discussed in this introduction is *accident.* While the average person considers accidents to be events that occur that are beyond an individual's control, safety professionals look at accidents in a more systematic and determined manner. Accidents are unplanned events that often result in injuries or damage that interrupt the routine operation of an activity. Accidents are always preceded by the unsafe act of employees or hazardous conditions in the workplace. Safety professionals know that accidents are caused by specific events that expose hazards and result in an incident. By taking appropriate action, most accidents can be eliminated. Chapter 5, examining accident causation, will pursue this topic in greater detail.

JOB TITLES OF INDIVIDUALS PERFORMING OCCUPATIONAL SAFETY AND HEALTH ACTIVITIES

There are many titles given to individuals who perform occupational safety and health activities. The following list describes just a few of those titles:

Safety Professional: Individuals who, by virtue of their specialized knowledge and skill and /or educational accomplishments, have achieved professional status in the safety field. They may also have earned the status of Certified Safety Professional (CSP) from the Board of Certified Safety Professionals.

Safety Engineer: Individuals who, through education, licensing, and/or experience, devote most or all of their employment time to the application of scientific principles and methods for the control and modification of the workplace and other environments to achieve optimum protection for both people and property.

Safety Manager: The individual responsible for establishing and maintaining the safety organization and its activities in an enterprise. Typically, the safety manager administers the safety program and manages subordinates including the fire prevention coordinator, industrial hygienist, safety specialists, and security personnel.

Industrial Hygienist: Although basically trained in engineering, physics, chemistry, or biology, this individual has acquired by study and experience a knowledge of the effects upon health of chemical and physical agents under various levels of exposure. The industrial hygienist is involved in the monitoring and analytical methods required to detect the extent of exposure, and the engineering and other methods used for hazard control.

Risk Manager: The Risk Manager in an organization is typically responsible for insurance programs and other activities that minimize losses resulting from fire, accidents, and other natural and man-made losses.

SAFETY AND HEALTH PROFESSIONAL'S ROLE AND RESPONSIBILITY

The specific roles and responsibilities of safety professionals depend upon the jobs that they are employed in or the types of hazards that are present at the companies where they work. Research (Kohn, 1991) examining the roles and responsibilities of safety professionals identified the following activities as those most frequently performed:

Hazard Recognition: identifying conditions or actions that may cause injury, illness, or property damage.

Inspections/Audits: evaluating/assessing safety and health risks associated with equipment, materials, processes, or activities.

Fire Protection: eliminating or minimizing fire hazards by inspection, layout of facilities, and design of fire suppression systems.

Regulatory Compliance: ensuring that all mandatory safety and health standards are satisfied.

Health Hazard Control: recognizing, evaluating, and controlling hazards that can create undesirable health effects, including noise, chemical exposures, radiation, or biological hazards.

Ergonomics: designing or modifying the workplace based upon an understanding of human physiological/psychological characteristics, abilities, and limitations.

Hazardous Materials Management: ensuring that dangerous chemicals and other products are stored and used in such a manner as to prevent accidents, fires, and the exposure of people to these substances.

Environmental Protection: recognizing, evaluating, and controlling hazards that can lead to undesirable releases of harmful substances into air, water, or the soil.

Training: providing employees with the knowledge and skills necessary to recognize hazards and perform their jobs safely and effectively.

Accident Investigation: determining the facts and causes related to an accident based upon witness interviews and site inspections.

Recordkeeping: maintaining safety and health information to meet government requirements, as well as to provide data for problem solving and decision-making.

Emergency Response Teams: organizing, training, and coordinating skilled employees to react to emergencies such as fires, accidents, or other disasters.

While this list provides examples of specific activities performed by the safety professional, the American Society of Safety Engineers (ASSE) has published a document titled *Scope and Functions of the Professional Safety Position*. This ASSE publication presents the "big picture" of the safety professional's roles and responsibilities (refer to the contents of this document below).

Figure 1-1. ASSE *Scope and Functions of the Professional Safety Position*.

Scope and Functions of the Professional Safety Position
American Society of Safety Engineers

The Scope of the Professional Safety Position

To perform their professional functions, safety professionals must have education, training and experience in a common body of knowledge. Safety professionals need to have a fundamental knowledge of physics, chemistry, biology, physiology, statistics, mathematics, computer science, engineering mechanics, industrial processes, business, communication, and psychology. Professional safety studies include industrial hygiene and toxicology; design of engineering hazard controls; fire protection; ergonomics; system and process safety; safety and health program management; accident investigation and analysis; product safety; construction safety; education and training methods; measurement of safety performance; human behavior; environmental safety and health; and safety, health, and environmental laws, regulations, and standards. Many safety professionals have backgrounds or advanced study in other disciplines, such as management and business administration, engineering, education, physical and social sciences, and other fields. Others have advanced study in safety. This extends their expertise beyond the basics of the safety profession.

Because safety is an element in all human endeavors, safety professionals perform their functions in a variety of contexts in both public and private sectors, often employing specialized knowledge and skills. Typical settings are manufacturing, insurance, risk management, government, education, consulting, construction, health care, engineering and design, waste management, petroleum, facilitates management, retail, transportation, and utilities. Within these contexts,

---

safety professionals must adapt their functions to fit the mission, operations, and climate of their employer.

Not only must safety professionals acquire the knowledge and skill to perform their functions effectively in their employment context, but also through continuing education and training they stay current with new technologies; changes in laws and regulations, and changes in the workforce, workplace, and world business, political, and social climate.

As part of their positions, safety professionals must plan for and manage resources and funds related to their functions. They may be responsible for supervising a diverse staff of professionals.

By acquiring the knowledge and skills of the profession, developing the mind set and wisdom to act responsibly in the employment context, and keeping up with changes that affect the safety profession, the safety professional is able to perform required safety professional functions with confidence, competence, and respected authority.

Functions of the Professional Safety Position

The major areas relating to the protection of people, property, and the environment are:

A. Anticipate, identify, and evaluate hazardous conditions and practices.

1. Developing methods for

a. anticipating and predicting hazards from experience, historical data, and other information sources.

b. identifying and recognizing hazards in existing or future systems, equipment, products, software, facilities, processes, operations, and procedures during their expected life.

c. evaluating and assessing the probability and severity of loss events and accidents which may result from actual or potential hazards.

2. Applying these methods and conducting hazard analyses and interpreting results.

3. Reviewing, with the assistance of specialists where needed, entire systems, processes, and operations for failure modes; causes and effects of the entire system, process, or operation; and any sub-systems or components due to

a. system, sub-system, or component failures.

b. human error.

c. incomplete or faulty decision making, judgments, or administrative actions.

d. weaknesses in proposed or existing policies, directives, objectives, or practices.

4. Reviewing, compiling, analyzing, and interpreting data from accident and loss event reports, and other sources regarding injuries, illnesses, property damage, environmental effects, or public impacts to

a. identify causes, trends, and relationship.

b. ensure completeness, accuracy and validity, or required information.

c. evaluate the effectiveness of classification schemes and data collection methods.

d. initiate investigations.

5. Providing advice and counsel about compliance with safety, health, and environmental laws, codes, regulations, and standards.

6. Conducting research studies of existing or potential safety and health problems and issues.

7. Determining the need for surveys and appraisals that help identify conditions or practices affecting safety and health, including those which require the services of specialists, such as physicians, health physicists, industrial hygienists, fire protection engineers, design and process engineers, ergonomists, risk managers, environmental professionals, psychologists, and others.

8. Assessing environments, tasks, and other elements to ensure that physiological and psychological capabilities, capacities, and limits of humans are not exceeded.

B. Develop hazard control designs, methods, procedures, and programs.

1. Formulating and prescribing engineering or administrative controls, preferably before exposures, accidents, and loss events occur to

a. eliminate hazards and causes of exposures, accidents, and loss event.

b. reduce the probability or severity of injuries, illnesses, losses, or environmental damage from potential exposures, accidents, and loss events when hazards cannot be eliminated.

2. Developing methods which integrate safety performance into the goals, operations, and productivity of organizations and their management and into systems, processes, and operations or their components.

3. Developing safety, health, and environmental policies, procedures, codes, and standards for integration into operational policies of organizations, unit operations, purchasing, and contracting.

4. Consulting with and advising individuals and participating on teams

a. engaged in planning, design, development, and installation or implementation of systems or programs involving hazard controls.

b. engaged in planning, design, development, fabrication, testing, packaging, and distribution of products or services regarding safety requirements and application of safety principles that will maximize product safety.

5. Advising and assisting human resources specialists when applying hazard analysis results or dealing with the capabilities and limitations of personnel.

6. Staying current with technological developments, laws, regulations, standards, codes, products, methods, and practices related to hazard controls.

C. Implement, administer, and advise others on hazard controls and hazard control programs

1. Preparing reports that communicate valid and comprehensive recommendations for hazard controls which are based on analysis and interpretation of accident, exposure, loss event, and other data.

2. Using written and graphic materials, presentations, and other communication media to recommend hazard controls and hazard control policies, procedures, and programs to decision-making personnel.

3. Directing or assisting in planning and developing educational and training materials or courses. Conducting or assisting with courses related to designs, policies, procedures, and programs involving hazard recognition and control.

4. Advising others about hazards, hazard controls, relative risk, and related safety matters when they are communicating with the media, community, and public.

5. Managing and implementing hazard controls and hazard control programs that are within the duties of the individual's professional safety position.

D. Measure, audit, and evaluate the effectiveness of hazard controls and hazard control programs.

1. Establishing and implementing techniques; which involve risk analysis, cost, cost-benefit analysis, work sampling, loss rate and similar methodologies; for periodic and systematic evaluation of hazard control and hazard control program effectiveness.

2. Developing methods to evaluate the costs and effectiveness of hazard controls and programs and measure the contribution of components of systems, organizations, processes, and operations toward the overall effectiveness.

3. Providing results of evaluation assessments, including recommended adjustments and changes to hazard controls or hazard control programs, to individuals or organizations responsible for their management and implementation.

4. Directing, developing, or helping to develop management accountability and audit programs which assess safety performance of entire systems, organizations,

processes, and operations or their components and involve both deterrents and incentives.

CONCLUSION

The modern occupational safety and health profession has roots that go to the beginnings of society. Concerns regarding the protection of human health, safety, and property are what form the basis for the profession. The following chapters of this textbook will address, in greater detail, the basic knowledge and skill area, critical for the successful implementation of a sound occupational safety management program.

Questions

1. Why does it make "good business sense" to have a good safety program? List four reasons.

2. Do you think most individuals are concerned with occupational safety and health issues? Why?

3. Why is it useful to study occupational safety and health historical events?

4. What is your definition of the term safety? How does it differ from the professional definition of this term?

5. What potential losses can result from safety and health hazards in the workplace?

6. What are some of the responsibilities of safety professionals?

References

Abercrombie, S. A. 1981. *Dictionary of terms used in the safety profession*. Park Ridge, IL: American Society of Safety Engineers.

American Society of Safety Engineers.1994. *Scope and functions of the professional safety position*. [Brochure]. Des Plaines, IL: American Society of Safety Engineers.

DeReamer, R. 1980. *Modern safety and health technology*. New York: Wiley & Sons, Inc.

Grimaldi, J. V., and Simonds, R. H. 1975. *Safety management*. 3rd ed. Homewood, IL: Richard D. Irwin, Inc.

Kohn, J.P., Timmons, D.L., and Besesi, M. "Occupational Health and Safety Professionals: Who Are We? What Do We Do?" *Professional Safety* 36:1.

National Safety Council. 1994. *Accident facts, 1994 edition*. Itasca, IL: National Safety Council.

Chapter 2

Safety Legislation

CHAPTER OBJECTIVES

After completing this chapter, you will be able to

- Explain the history of safety and health legislation.
- Understand the Occupational Safety and Health Act of 1970.
- Identify the origins of OSHA standards.
- Know the specific requirements of the Act.
- Understand the OSHA inspections and resultant actions.

CASE STUDY

When Smith's machine shop was inspected, Eric Smith was shocked. With only five employees, Eric thought he was exempt from OSHA inspections. He was even more surprised to learn that one of his employees had complained. After a careful investigation, Eric learned the name of the complainant and nearly fired him. Eric's attorney informed him that retaliatory action against an employee for exercising his rights under OSHA regulations is illegal.

LEGISLATIVE HISTORY

The history of occupational safety and health has been dominated by legislation. Governments have observed problems and have attempted to solve those problems through the enactment of laws. In this country, the regulations culminated with the passage of the Occupational Safety and Health Act in 1970.

The eye-for-an-eye principle dominated early attempts to legislate safety. The government effort to encourage safer work places first revolved around punishing the wrong doer. This concept pervaded Babylonian law over 4000 years ago and was a forerunner to the famous Code of Hammurabi which was written in 2100 BC, during the thirtieth year of his reign. The Code required ship builders to repair defects of construction and damage caused by those defects for one year following delivery. Ship captains were required to replace goods lost at sea and to pay a fine equal to half the value of any lost ships which were refloated. If a slave was injured by someone other than the master, the master received compensation for the loss. Carelessness and neglect were considered unacceptable for skilled workers and professionals. Losses caused by errors on the part of these early professionals were punished using the eye-for-an-eye concept which is also found in the Old Testament. A physician whose mistakes lead to the loss of a citizen's life could find his hands cut off, or a builder could have his own child killed if his shoddy work lead to the loss of another's child. From the time of Hammurabi and his contemporaries, little is known about attempts to legislate safety until after the Dark Ages. The British developed its laws out of concern for its children (Grimaldi, 1989).

A review of events of the 1800s reveals a number of tragedies that met with little government response. As the media coverage became more widespread and the early information network became accessible through books, newspapers, and magazines, American citizens began to expect and demand that their government provide protection from employers. Coinciding with this increased public awareness, a series of events marks the early development of safety and health legislation in the United States.

This reactive pattern continued as concern for occupational safety and health began to emerge. As public outcry to disasters or outrageous actions on the part of employers became strong enough to overcome the influence exerted by employers on early legislatures, laws were enacted to protect the workers. Early industrial plants were little more than a series of traps consisting of open machinery and unguarded moving equipment which could maim or kill a worker in seconds. The following review of significant events in the development of safety and health will therefore be useful:

In 1903, a fire roared through Chicago's Iroquois Theater killing 602 people. The fire engulfed the stage almost immediately. As the audience panicked, flames swept across the perimeter of the auditorium and finally to the seats themselves. Fire exits were few and poorly marked. The iron pillars became red hot and eventually melted.

In Monongah, a sleepy West Virginia town in the hills of Appalachia, 362 Coal miners were killed in 1907. Nearly every family in the town was affected by the disaster but little was done for the survivors.

The Pittsburgh Survey (1907-1908), a twelve-month study in Allegheny County, Pennsylvania was sponsored by the Russell Sage Foundation. It was found that there were 526 occupational fatalities in one year. Survivors and the workers themselves were forced to carry the cost of losses. When the bread winners were gone, there was no further compensation of any kind. Employer incentives to reduce safety and health risks for workers were called for.

In 1910, the U.S. Bureau of Mines was created by the Department of Interior to investigate the causes of mine accidents, study health hazards, and to find means for taking corrective action.

A fire gutted a new structure in the Triangle Shirtwaist Factory in New York City. This 1911 tragedy killed 145 workers whose escape was prevented due to the employer's locking of the exits against theft.

In 1911, Wisconsin passes the first successful workers' compensation plan in the U.S.

The first National Safety Congress convened in 1912. It led to the formation of the National Safety Council in 1913.

The Safety to Life Committee formed by the National Fire Protection Association in 1913, eventually led to the development of the *Life Safety Code*.

In 1935, Roosevelt's New Deal included the passage of legislation that mandated a forty-hour work week.

The Walsh-Healey Public Contracts Act of 1936 banned hazardous work done under federal contracts over $10,000. This act was a forerunner of the Occupational Safety and Health Act.

In 1968, seventy-eight coal miners were killed in Farmington, West Virginia, when explosion ripped through the Consolidated Coal Company Mine. Only a few miles from the infamous Monongah site, this disaster devastated the whole community. A town, a state, and a nation were outraged and called for federal intervention into the conditions killing citizens in the workplace.

The Coal Mine Health and Safety Act of 1969 was passed 72-0 in the U.S. Senate. This established the Mine Enforcement Safety Administration (MESA) which later became the Mine Safety and Health Administration (MSHA). MSHA governs the safety within the coal mines as OSHA does for general industry and construction (discussed below).

In 1970, the Williams-Steiger (Occupational Safety and Health) Act was passed establishing the Occupational Safety and Health Administration. Prior to the establishment of OSHA, the responsibility for occupational safety and health rested primarily with state governments.

OCCUPATIONAL SAFETY AND HEALTH ACT

Although more than 90 million Americans spent their days on the job, until 1970, no uniform and comprehensive provisions existed for their protection against workplace safety and health hazards. In 1970, Congress considered annual figures such as these:

- Job-related accidents accounted for more than 14,000 worker deaths.
- Nearly 2.2 million workers were disabled.
- Ten times as many person-days were lost from job-related disabilities.
- Estimated new cases of occupational diseases totaled 300,000.

The Occupational Safety and Health Act (OSHAct) of 1970 was passed by Congress "...to assure so far as possible every working man and woman in the Nation safe and healthful working conditions and to preserve our human resources" (U.S. DOL, 1991). Under the Act, the Occupational Safety and Health Administration (OSHA) was created within the Department of Labor to:

- Encourage employers and employees to reduce workplace hazards and to implement new or improve existing safety and health programs
- Provide for research in occupational safety and health to develop innovative ways of dealing with occupational safety and health problems
- Establish "separate but dependent responsibilities and rights" for employers and employees for the achievement of better safety and health conditions (U.S. DOL, 1991). Maintain a reporting and recordkeeping system to monitor job-related injuries and illnesses
- Establish training programs to increase the number and competence of occupational safety and health personnel
- Develop mandatory job safety and health standards and enforce them effectively
- Provide for the development, analysis, evaluation, and approval of state occupational safety and health programs (OSHATI, 1994, chap. 1)

Who is Covered?

In general, the coverage of the Act extends to all employers and their employees in the 50 states, the District of Columbia, Puerto Rico, and all other territories under federal government jurisdiction. Coverage is provided either directly by federal OSHA or through an OSHA-approved state program. As defined by the Act, an employer is any "person engaged in a business affecting commerce who has employees, but does not include the United States or any State or political subdivision of a State" (U.S. DOL, 1991, p. 5).

The following are not covered under the Act:

- Self-employed persons
- Farms at which only immediate members of the farm employer's family are employed
- Workplaces already protected by other federal agencies under other federal statutes (OSHATI, 1994, chap. 1)

Under the Act, federal agency heads are also responsible for providing safe and healthful working conditions for their employees. The Act requires agencies to comply with standards consistent with those OSHA issues for private sector employers. OSHA conducts federal workplace inspections in response to employees' reports of hazards and as part of a special program which identifies federal workplaces with higher-than-average rates of injuries and illnesses (OSHATI, 1994, chap. 1).

OSHA cannot fine another federal agency for failure to comply with OSHA standards, and it does not have authority to protect federal employee "whistleblowers." Federal employee whistle blowers are protected under the Whistleblower Protection Act of 1989 (OSHATI, 1994, chap. 1).

OSHA provisions do not apply to state or local governments' employees. The Act requires that states desiring approval to maintain their own programs must provide programs that cover state and local government workers, and they must be at least as effective as their programs for private employees. State plans may also cover only public sector employees.

Employers of 11 or more individuals are required to maintain records of occupational injuries and illnesses. OSHA recordkeeping is not required for certain retail trades and some service industries. Employers exempt from recordkeeping must comply with the other OSHA standards, including displaying the OSHA poster.

OSHA STANDARDS

OSHA standards fall into four categories: General Industry, Maritime, Construction, and Agriculture. The standards are available in the following volumes:

Volume I   General Industry Standards and Interpretations (includes Agriculture)

Volume II Maritime Standards

Volume III Construction Standards

Volume IV Other Regulations and Procedures

Volume V Field Operations Manual

Volume VI OSHA Technical Manual (OSHATI, 1994, chap. 1)

These are available from the Superintendent of Documents, U.S. Government Printing Office, Washington, DC 20402 and commercial publishers. Because some states adopt and enforce their own standards, copies of those may be obtained from the individual states.

The Act consists of twenty-one sections as outlined in Appendix A.

Origin of OSHA Standards

Initially, the OSHA standards were taken from three sources: consensus standards, proprietary standards, and federal laws in effect when the Act became law (OSHATI, 1994, chap. 2).

Consensus standards are developed by industry-wide standard-developing organizations which are discussed and substantially agreed upon through consensus by industry. OSHA has incorporated the standards of the two primary standards groups, the American National Standards Institute (ANSI) and the National Fire Protection Association (NFPA). For example, ANSI Standard B56.1-1969, Standard for Powered Industrial Trucks, covers the safety requirements relating to the elements of design, operation and maintenance of powered industrial trucks (OSHATI, 1994, chap. 2).

Proprietary standards are prepared by professional experts within specific industries, professional societies, and associations. These proprietary standards are determined by membership vote, as opposed to consensus. An example of these would be the Compressed Gas Association's, *Pamphlet P-1, Safe Handling of Compressed Gases*. This proprietary standard covers requirements for the handling, storage, and use of compressed gas cylinders (OSHATI, 1994, chap. 2).

Some preexisting federal laws are also enforced by OSHA, including the Federal Supply Contracts Act (Walsh Healey) and the Contract Work Hours and Safety Standards Act (Construction Safety Act). Standards issued under these acts are now enforced in all industries where they apply (OSHATI, 1994, chap. 2).

When the OSHAct was first passed, much criticism stemmed from the fact that the legislation was a hodgepodge of rules of thumb and guidelines which were never intended to be made into laws. Since the passage of the Act, many of the trivial regulations have been changed or eliminated in an attempt to make it a more reasonable standard for workplace performance.

Horizontal and Vertical Standards

Standards are referred to as being either *horizontal* or *vertical* in their application. Most standards are *horizontal* or *general* in that they apply to any employer. Standards relating to fire protection or first aid are examples of horizontal standards (OSHATI, 1994, chap. 2).

Some standards are relevant only to a particular industry, and are called *vertical* or *particular* standards. Examples are standards that apply to the longshoring or construction industry.

FINDING THE OSH ACT

The Occupational Safety and Health Act appears in the Code of Federal Regulations (CFR) 29. It is divided into the subparts noted in Appendices A and B. Each subpart addresses a different major topical area. These areas include such items as Subpart D-Walking-Working Surfaces. This subpart appears in 1910.21-1910.32. Subpart E-Means of Egress follows in 1910.35-1910.70. Each subpart is further broken down so that when one looks under Subpart E, one will find the subsection, 1910.37, Means of egress, general. The reader can turn to that section to see an explanation of means of egress and find that it is broken down even further. References to specific standards are typically found as follows:

29 CFR 1910.110(b)(13)(ii)*(b)(7)(iii)* indicates that the specific standard in question appears in

- Title-29
- Code of Federal Regulations-CFR

- Part-29
- Section-.110

Subsections appear first as lower case letters, numbers, and Roman numerals. Subsections of subsections appear as italicized, lower-case letters, numbers, and Roman numerals (OSHATI, 1994, chap. 2).

Most universities and many public libraries carry copies of the Code of Federal Regulations. They are also available from the Superintendent of Documents in Washington and from Government Institutes, Inc.

SPECIFIC REQUIREMENTS OF THE ACT

Employers are responsible for knowing the standards applicable to their establishments. When an OSHA inspection is performed, the assumption made is that the employer is aware of the law and has already attempted to comply with it. Any violations are subject to corrective legal action, typically consisting of fines. Employees must also comply with all rules and regulations that are applicable to their own actions and conduct. It is the employer's responsibility to assure employee compliance.

Where OSHA does not have specific standards, employers are responsible for following the Act's general duty clause (Section 5(a)(1)). The general duty clause requires that every working person must be provided with a safe and healthful workplace. It specifically states: "Each employer shall furnish to each of his employees employment and a place of employment which is free from recognized hazards that are causing or are likely to cause death or serious physical harm to his employees" (OSHA Standards, 1993). *Recognized hazard* is one that is detectable by the senses or instrumentation, of which there is common knowledge, that is discoverable under usual inspection practices, or of which the employer has knowledge. Incidentally, if the employer has knowledge, an inspector will consider the hazard to be a willful violation which carries an increased penalty.

The general duty clause extends OSHA's authority beyond the specific requirements of the standards when a recognized workplace hazard exists or potentially exists. In 1990-1991 alone, "the General Duty Clause was used as the basis for over 2,232 OSHA citations. Over $500,000 in penalties were collected as a result of these citations. Only nine other sections of the OSHA standards (Parts 1910 and 1926) are cited more frequently as the basis for OSHA violations than Section 5(a)(l)" (OSHA Standards, 1992). The general duty clause is often used by OSHA when there is no specific standard that applies to a recognized hazard in the workplace. OSHA may also use the General Duty Clause when a standard exists, but it is clear that the hazards involved warrant additional precautions beyond what the current safety standards require.

EMPLOYER RESPONSIBILITIES AND RIGHTS

Aside from providing a workplace free from recognized hazards likely to cause death or serious physical harm, the employer has other responsibilities. An employer must

- Examine workplace conditions to make sure they comply with applicable standards.
- Minimize or reduce hazards.
- Use color codes, posters, labels, or signs when needed to warn employees of potential hazards.
- Provide training required by OSHA standards.
- Keep OSHA required records.
- Provide access to employee medical records and exposure records to employees or their authorized representatives.

Employers have the right to seek advice and off-site consultation as needed by writing, calling, or visiting the nearest OSHA office (OSHATI, 1994, Chapter 2).

When an inspection visit occurs, the employer must

- Be advised by the compliance officer of the reason for the inspection.
- Accompany the compliance officer on the inspection.
- Be assured of the confidentiality of any trade secrets observed by an OSHA compliance officer during an inspection

Although OSHA does not cite employees for violations of their responsibilities, each worker is required to comply with all occupational safety and health standards that are applicable. Employees also have the right to ask for safety and health on the job without fear of punishment. If employees are discriminated against for exercising their rights under the Act, OSHA may take the employer to court with no expense to the employees.

INSPECTIONS

To enforce its standards, OSHA is authorized under the Act to conduct workplace inspections. Every establishment covered by the Act is subject to inspection by OSHA compliance safety and health officers who are chosen for their knowledge and experience in the occupational safety and health field. Compliance officers are vigorously trained in OSHA standards and in the recognition of safety and health hazards. Inspections occur as a result of the following priorities established by OSHA:

Imminent danger situations are inspected first. Where there is reasonable certainty that an employee is exposed to a hazard likely to cause death or immediate serious physical harm, OSHA will try to respond as soon as possible. OSHA may become alerted to these conditions by a complaint or other means. Inspectors may see a story on the evening news or may even be driving by a job site where they see an imminent danger situation.

Catastrophes and fatal accidents are investigated next. If three or more employees are hospitalized or if an employee is killed, OSHA must be notified within eight hours.

Employee complaints, alleging violation of standards or of unsafe or unhealthy working conditions, are investigated next. The employee has the right to remain anonymous to his employer.

Programmed high-hazard inspections are given the next priority. These are aimed at specific high-hazard industries, occupations, or health substances. Selection is based on factors such as death, injury, and illness incidence rates and employee exposure to hazardous substances.

Follow-up inspections are given last priority. These are used to determine if previously cited violations have been corrected (OSHATI, 1994, chap. 1).

INSPECTION PROCESS

Under the Act, "upon presenting appropriate credentials to the owner, operator or agent in charge," (U.S. DOL, 1991) an OSHA compliance officer is authorized to

- Enter without delay and at reasonable times any factory, plant, establishment, construction site or other areas, workplace, or environment where work is performed by an employee of an employer;" (U.S. DOL, 1991) and to
- Inspect and investigate during regular working hours, and at other reasonable times, and within reasonable limits and in a reasonable manner, any such place of employment and all pertinent conditions, structures, machines, apparatus, devices, equipment, and materials therein, and to question privately any such employer, owner, operator, agent, or employee (U.S. DOL, 1991).

Inspections are generally conducted without advance notice. In fact, alerting an employer in advance of an OSHA inspection can bring a criminal fine of up to $1,000 and/or a six-month jail term. If an employer refuses to admit an OSHA compliance officer, or if an employer attempts to interfere with the inspection, the Act permits appropriate legal action.

Once credentials are presented, the compliance officer will explain in an *opening conference* why the inspection is being performed, the scope of the inspection, and the standards that apply. An authorized employee representative is given the opportunity to attend and to accompany the compliance officer on the inspection.

Following the conference, the compliance officer will inspect the OSHA records including the OSHA 200. The inspector may also request copies of other required records such as the hazard communication or lockout/tagout programs.

An *inspection tour* then takes place. The officer may talk to employees in private about safety and health conditions, as well as practices in their workplace. An employer's representative should accompany the officer and keep a careful record of everything inspected. Any comments made by the inspector should be recorded. Split samples should be requested anytime samples are taken. Copies of any photographs taken should be requested.

A *closing conference* occurs near the end of the visit. During this time, the employer or an employer's representative should ask questions in order to clearly understand any violations recorded by the compliance officer. The employees' representative may be present during this conference. No penalties will be assigned at this time.

CITATIONS AND PENALTIES

After the compliance officer reports findings, the area director determines what citations, if any, will be issued, and what penalties, if any, will be proposed. Citations inform the employer and employees of the regulations and standards alleged to have been violated and of the proposed length of time set for their abatement. The employer receives the citations and notices of proposed penalties by certified mail within six months of the inspection. The employer must post a copy of each citation at or near the place a violation occurred, for three days or until the violation is abated, whichever is longer.

OSHA may issue any of four types of citations:

Citations for *willful violations* are issued when the employer disobeys, with an intentional disregard of, or plain indifference to, the requirements of the OSHAct and regulations. These can be assessed if the employer was aware that a hazardous condition existed and made no reasonable effort to eliminate the condition. The employer need not be guilty of malicious intent to be considered willful. Under OSHA's penalty structure, the maximum penalty for a willful violation is now $70,000. The minimum penalty is $10,000. Criminal charges can be brought against an employer if an employee fatality is caused by such negligence.

Citations for *serious* violations are issued when there is a substantial probability that death or serious physical harm could result and that the employer knew or should have known of the hazard. Violations of the general duty clause are considered serious. The maximum penalty is now $7,000.

Citations for other than serious violations are issued when a situation would affect safety or health but there is a small probability of the hazard resulting in death or serious physical harm. There is often no penalty assessed, but the hazard must still be corrected. If there is a high probability of the hazard resulting in an injury or illness, then the maximum penalty is $1,000. The OSHA Regional Administrators have the authority to impose a penalty of up to $7,000 if the circumstances warrant.

Regulatory citations are issued for

- No OSHA poster-$1,000
- No OSHA 200 Log-$1,000
- Failure to post citations-$3,000
- Failure to report within 8 hours a fatality or accident which hospitalizes 3 or more employees-$5,000

Repeat violation citations are issued when the original violation has been abated, but upon reinspection, another violation of the previously cited section of a standard is noted. They may be inadvertent, but if they are found to be willful, both a willful and a repeat citation may be issued. For a first repeat violation, penalties assessed are multiplied by a factor of 2 for employers with less than 251 employees and by 5 for larger employers. The multiplier goes to 5 for a second repeat offense for small employers and 10 for large. OSHA Regional Administrators have the authority to use a multiplication factor of up to 10 for small employers in order to achieve the necessary deterrent effect. Failure to abate within the prescribed period can result in a penalty for each day of the violation beyond the abatement date.

Under OSHA's egregious policy, each fine may be multiplied by the number of employees exposed or by the number of times the violation occurs in the workplace. It is important to note that the Act also authorizes criminal penalties for certain violations. Penalties may be reduced for employers with less than 251 employees, good faith efforts, or a good safety record in the last three years.

APPEALS PROCESS

Within 15 working days of the receipt of the citation, an employer may submit a written objection to OSHA. Once an inspection takes place, those assigned to receive the citation in the mail for a company need to watch for and turn over any correspondence from OSHA. Companies that ignore the fifteen-day deadline or for some other reason fail to meet it, will find they have also missed their right to appeal. The employer may contest a citation, a penalty, and/or an abatement date. An OSHA area director has the authority to make adjustments based on objections. If no agreement can be reached with the director, the objection may be forwarded to the Occupational Safety and Health Review Commission, which operates independently of OSHA. Appeals beyond the Commission go through the appeals courts.

OSHA-APPROVED STATE PROGRAMS

The Act encourages states to develop and operate, under OSHA guidance, state job safety and health plans. Once a state plan is approved, OSHA funds up to 50 percent of the operating costs of the program. State plans are required to provide standards and enforcement programs, as well as voluntary compliance activities which are at least as effective as the federal program. They must also provide coverage for state and local government employees. In addition, OSHA permits states to develop plans limited to public sector coverage (state and local government). In such cases, private sector employment remains under federal jurisdiction.

STANDARDS DEVELOPMENT

Once OSHA has developed plans to propose, amend or delete a standard, it publishes these intentions in the *Federal Register* as a "Notice of Proposed Rulemaking," (U.S. DOL, 1991) or often as an earlier "Advance Notice of Proposed Rulemaking" (U.S. DOL, 1991). The Advance Notice (U.S. DOL, 1991) is used to solicit information that can be used in drafting a proposal. A notice of proposed rulemaking will include the terms of the new rule and provide a specific time (at least 30 days from the date of publication, usually 60 days or more) for the public to respond.

Interested parties who submit written arguments and pertinent evidence may request a public hearing on the proposal when none has been announced in the notice. When such a hearing is requested, OSHA will schedule one, and will publish, in advance, the time and location in the *Federal Register*.

After the close of the comment period and public hearing, if one is heard, OSHA must publish in the *Federal Register* the full, final text of any standard amended or adopted and the date it becomes effective, along with an explanation of the standard and the reasons for implementing it. OSHA may also publish a determination that no standard or amendment needs to be issued.

OTHER CONSIDERATIONS

Since the inception of OSHA, the construction industry has been underrepresented. According to one OSHA administrator, construction has always been an afterthought at OSHA. Much of what has been legislated in construction has been after the fact. Once rules were passed for general industry, the question was asked, "Now what do we do with construction?" In recent years, this policy has begun to change. Construction has been a hazardous industry and those hazards have been recognized by OSHA. Construction standards appear in 29 CFR 1926. OSHA has *target programs* that give a high priority to inspection of certain hazards. Trenching and shoring is an area that has been recently targeted by OSHA. Contractors using these processes will find themselves subjects of OSHA inspections and fines if problems persist.

NIOSH AND OSHRC

The OSHAct also created the National Institute of Occupational Safety and Health (NIOSH) and the Occupational Safety and Health Review Commission (OSHRC). NIOSH operates within the Department of Health and Human Services (HHS) under the Center for Disease Control (CDC) to develop occupational safety and health standards for recommendation to the Secretary of Labor and the Secretary of HHS, and to fulfill the research and training functions of the secretary of HHS. It is headquartered in Washington, DC, but carries out many of its functions at its facilities in Cincinnati, Ohio, and Morgantown, West Virginia. It also works through contracts with fourteen Educational Research Centers and over forty Project Training Grantees around the country. These institutions provide research and training in safety and industrial hygiene on behalf of NIOSH.

The Occupational Safety and Health Review Commission is an independent and autonomous, quasi-judicial board charged with hearing cases on appeal from OSHA. Its three members are appointed by the President with the advice and consent of the senate. It meets periodically to review cases on appeal. Once appeals are heard by the commission, they go through the federal appeals court system.

FUTURE TRENDS

The agricultural industry has not fared as well as other industries. Although agricultural problems are comparable to those in the construction or mining industries, they have not had much attention from OSHA. Many agricultural operations are small, family-owned businesses that manage to escape OSHA's attention. If no employees outside the family work in the operation, it is automatically exempt. However, many farmers are unaware of the regulations to report incidents and illnesses, so their operations go largely unnoticed. As a better result of recordkeeping, and as farmers become more aware of the steps they can take to reduce workplace incidents, OSHA will undoubtedly play a more important role in the agricultural industry.

As the workplace continues to become more complex, and as OSHA finds that higher numbers of workers are killed or injured in different ways than in the past, the emphasis will shift. Homicide was the second leading cause of death in 1993, accounting for 17 percent of fatal injuries to workers. The leading cause of death was automobile accidents. OSHA may begin exploring these areas and attempt to generate creative methods of targeting and reducing incidents that, historically, have not been considered a workplace problem. Fast food merchandisers, delivery services, and nighttime retailers may be in line for future inspections as workplace homicide continues to become more of a concern for OSHA.

Although there has been discussion on changing the way OSHA enforces compliance through regulations, companies will continue to be required to comply with applicable standards and codes. Future directives may force OSHA to target high- incidence industries based on a program similar to the Maine 200 program. In that program, companies with the highest workers' compensation are targeted for inspection. Additional efforts may also be placed on targeting and penalizing companies that do not have the basics in place. Compliance officers may only check a few critical areas. If these "look good" or if the company is in compliance, they will move on to the next inspection. If they do not look good or the company is out of compliance, then a complete inspection will take place.

CONCLUSION

The Occupational Safety and Health Act was the culmination of centuries of governmental response to occupational safety and health problems. Since 1970, a number of additional regulations designed to protect workers have been passed. These will be addressed in other chapters. As was noted, incident rates have declined and this decline appears to be a result of legislation. It is important to recognize, however, that conformance to legislation is not enough to create a safe and healthful working environment. Fortunately, compliance requirements have encouraged a safer and healthier workplace and helped assist otherwise unsafe companies toward preserving the health and safety of their workers.

Questions

1. Can safety and health be legislated? What are the limitations of safety legislation? Where would we be without it?

2. How do you think the future of safety legislation would be affected by another major occupational disaster of the magnitude of a Monongah mine disaster? Do you think passage of new safety and health legislation is still of a reactive nature?

3. Although safety and health fines have gone up significantly in recent years, do you think it matters how high they are when the likelihood of being inspected is so low? Discuss this with people in industry and note their opinions.

References

Grimaldi, J. V. and Simonds, R. H. 1989. *Safety management.* Homewood, IL: Richard D. Irwin.

Georgia Tech Research Institute. 1994, December, 12-16. *OSHA 501 training.* Atlanta, GA: Georgia Tech Research Institute.

OSHA Training Institute. 1994. *A guide to voluntary compliance in safety and health.* Atlanta, GA: Georgia Tech Research Institute.

United States Department of Labor. 1991. *All about OSHA.* Washington, DC: U.S Government Printing Office.

Chapter 3

Workers' Compensation and Recordkeeping

CHAPTER OBJECTIVES

After completing this chapter, you will be able to

- Explain the concept of workers' compensation.
- Describe the evolution of workers' compensation.
- Describe the different types of workers' compensation claims.
- Explain the basis for workers' compensation rates.
- Identify the basic record keeping requirements.

CASE STUDY

Joe Derek often takes work home with him to complete some of the details that he is unable to finish at the office. Although his employer has never encouraged him to work at home, he is aware of the practice and thankful that the work gets done. One evening on the way home from work, Joe is injured in an automobile accident. His employer is surprised when Joe files for and is awarded workers' compensation.

EARLY WORKERS' COMPENSATION LAWS

Early efforts to compensate victims of workplace accidents stretch back to the Code of Hammurabi. In modern times, efforts have been made to force employers to pay for injuries that are suffered by employees while they are on the job. Modern efforts can be traced to Prussia, where in 1838, legislation was passed permitting railroad workers to collect for injuries they suffered on the job. In 1884, Otto Von Bismarck, enacted the first workers' compensation law for all German workers. Wisconsin passed the first successful workers' compensation law in the United States in 1911.

The workers' compensation laws were designed to compensate victims of workplace accidents by forcing the employer to pay the workers money for their injuries and time lost. Prior to the passage of these laws, the only recourse that a worker had was to sue the employer under civil law, a form of common law. Common law is unwritten laws based on judicial decisions of the past, as opposed to statutory law which is prepared and enacted by legislative bodies. In early common law suits, employers successfully utilized the following defenses:

1. Assumption of risk. The employees accepted the risks they were facing when they accepted the job. By doing so, they gave up any right to collect compensation for injuries.

2. Contributory negligence. Since the employees played a part, regardless of how small that part might have been, in the injury, they are not permitted to recover compensation for their injuries.

3. Fellow-servant rule. The employer is not at fault because the accident was the fault of another employee or other employees.

These three common law defenses were eliminated after workers' compensation laws were enacted by each state. Employers were sold on the laws because in order to collect, employees had to waive their rights to sue. The employee was glad to do so because now he was able to collect when he had previously been unable to do so. By 1948, every state had passed a workers' compensation law. Please note that when those laws were passed they were referred to as "work*men*'s compensation" laws. Today, they are referred to as *workers'* compensation laws.

Prior to the passage of the workers' compensation laws, the full cost of production was not borne by the employer but instead, was passed on to the employee. Rather than engineer a solution to a simple problem, such as providing a guard on an open sawblade, the employer would permit the employee to run the risk of being injured. It was cheaper to replace the employee than it was to guard the machines. If the employee were hurt, it was left to the relatives and neighbors to provide support.

Workers' compensation forced the employer to pay at least part of that support. In return, the employee agreed not to sue the employer for damages.

Most states prohibit the suing of businesses that comply with the Act except for deliberate assaults and conditions so flagrantly unsafe as to make injury virtually certain. This has worked as a win/win situation for a number of years. Recently, however, employees who have collected workers' compensation benefits have been successfully following up with lawsuits against the employer by using these exceptions as loopholes in the law. Following a successful suit, the employee simply repays the workers' compensation fund for money already received. This new practice is a result of the large awards to plaintiffs by sympathetic juries who often perceive employers and their insurance companies as bottomless sources of cash. Lawyers may engage clients on a no-fee basis and will not be paid unless an award is won. If they can demonstrate to a potential litigant that they won't be out any money if they lose but may stand to profit substantially if they win, then the employee may agree to enter into the suit. This relationship violates the spirit of the workers' compensation legislation but is a result of one of the inherent flaws in the system.

When a worker is injured and stands to collect benefits, that worker must first typically undergo a waiting period of from three to fourteen days. Benefits may or may not be paid retroactively to the first day of lost work and wages. Once the benefits begin, they are limited to two-thirds of the worker's wages or to a cap which can run as high as $1,500 per month or more depending on the state. Benefits extend for a limited period of time and a maximum amount of money. Because workers who are off from work for a period of time often suffer a decrease in their standard of living, litigation may appear to be an attractive option. If workers were permitted to receive benefits equal to their wages, however, there would be a strong incentive to not return to work. Even at two-thirds wages, some employees have been found to be malingering, so that they can receive benefits without working.

Injuries are categorized in one of the following ways:

Partial: when the employee can still work but is unable to perform all duties of the job due to the injury, as would often be the case with a broken finger or a severed toe.

Total: when the employee is unable to work or perform substantial duties on the job, as would often be the case with a severe back injury or blindness.

Temporary: when the employee is expected to fully recover, as would be the case with a broken limb or a sprain.

Permanent: when the employee will suffer the effects of the injury from now on, as would be the case with a severed limb, blindness, or permanent hearing loss.

Specific cases would be categorized as temporary partial, temporary total, permanent partial, or permanent total. Other categories of benefits include *retraining incentive benefits* for employees who may have specific injuries. These benefits are paid for a limited period of time to aid the injured in pursuing additional education or training. Vocational rehabilitation services may be offered to employees who are eligible for permanent total disability benefits who actively participate in a vocational rehabilitation program. *Survivors* of employees killed in industrial accidents may be entitled to their benefits as well.

Even businesses with only a few employees are required to carry workers' compensation insurance for their workers. Some states require companies to provide benefits even for one part-time employee. Large companies that may have the appropriate financial resources to cover anticipated losses to self insure are one of the few exceptions to this rule.

Who is covered by workers' compensation? The following is an example of the coverage that exists in one state:

- Every person, including a minor, whether lawfully or unlawfully employed, in the service of an employer
- Every executive officer of a corporation
- Every person in the service of the state, county, or city
- Every person who is a member of a volunteer ambulance service, fire, or police department
- Every person who is a regularly enrolled volunteer member or trainee of the civil defense corps

- Every person who is an active member of the National Guard
- Every person performing service in the course of the trade, business, profession, or occupation of an employer at the time of injury
- Every person regularly selling or distributing newspapers on the street or to customers at their homes or places of business
- Owner(s) of a business, whether or not employing any other person to perform a service for hire

Exemptions

The following employees are *exempt* from the coverage of workers' compensation:

- Any person employed as a domestic servant in a private home
- Any person employed for not exceeding twenty consecutive work days in or about the private home of the employer
- Any person performing services in return for aid or sustenance only, received from any religious or charitable organization
- Any person employed in agriculture
- Any person participating as a driver or passenger in a voluntary vanpool or carpool program while that person is on the way to or from his place of employment
- Any person who would otherwise be covered but elects not to be covered.

If an employee is injured during a work assignment outside of the state where he is employed, he is still eligible to collect his benefits. In some states, benefits are reduced if the employee is not using the safety equipment required by the employer at the time of the injury.

Premium Calculation

Rates that employers pay are based on the annual payroll in a given Standard Industrial Classification (SIC) code. SIC is the statistical classification used by the federal government to categorize business. The SIC for each business can be found in the *Standard Industrial*

Classification Manual published by the federal government. Workers' compensation is expressed in dollars per $100 of payroll. SIC codes that are considered to be higher risk because of higher numbers of injuries or deaths within those groups pay more.

Experience Modification

In addition to the classification rate, companies are assigned an *experience modification* based upon the claims activity of their specific business. It begins at zero and then changes either upward or downward as the claims activity of the business changes. It is possible that the experience modification can change yearly, but it typically takes into account the past three years.

Self-Insured

Employers may be self-insured if they are able to furnish to the workers' compensation board satisfactory proof of their financial ability to directly pay the compensation and if they deposit an acceptable security, indemnity, or bond (Champa, 1982).

RECORDKEEPING

Case Study

Jamie Smith is discouraged by the recordkeeping requirements of the company. She keeps two forms for OSHA and an additional form for workers' compensation requirements. When a compliance officer arrives for a routine inspection, she is surprised to learn that she does not have to keep all three forms. She is also surprised to learn that the form must be posted during the month of February. She has not posted the form in the past, but now assumes she should after the comment made by the compliance officer. The biggest surprise arrives three months later when the company receives a fine because the form was not signed by the appropriate company official. Unfortunately, the fact that no one verified the form cost the company $1,000.

Background

Before the Occupational Safety and Health Act of 1970 became effective, there was no centralized and systematic method for monitoring occupational safety and health problems. Statistics on job injuries and illnesses were collected by some states and by some private organizations. National figures were based on less-than-reliable projections. With the passage of OSHA came the first basis for consistent nationwide procedures, a vital requirement for gauging problems and solving them (U.S. DOL, 1991, p. 13).

Accurate injury and illness records are essential in providing information for your safety and health program. Because injury and illness recording is already a requirement under law (29 CFR 1904), you should use the information maintained on forms OSHA 200 and 101 to your advantage to

- reveal which operations are most hazardous;
- determine weaknesses in your safety and health program;
- judge the effectiveness of your program by comparing it with past records or records of other similar plans;
- aid in accident analysis and investigation;
- identify the causes of occupational diseases by relating them to particular exposures, or processes, or both; and
- satisfy legal and insurance requirements.

Accurate records can be used to analyze illnesses and injuries so problem areas can be identified and corrected" (NIOSH, 1979, p. 20).

Who Must Keep Records

Employers with eleven or more employees are subject to OSHA recordkeeping requirements as stated by 29 CFR 1904. They must maintain a record of occupational injuries and illnesses as they occur. Employers with ten or fewer employees are exempt from keeping such records unless selected by the Bureau of Labor Statistics (BLS) to participate in the Annual Survey of Occupational Injuries and Illnesses. They are, however, subject to the rest of the OSHA regulations applicable to their industry and jobs. The purpose of keeping records is to permit the BLS survey material to be compiled, to help define high hazard industries, and to inform employees of the status of their employer's record (U.S. DOL, 1991).

Occupational Injury

An "occupational injury" is any injury such as a cut, fracture, sprain, or amputation that results from a work-related accident or from exposure involving a single incident in the work environment.

Occupational Illness

An occupational illness is any abnormal condition or disorder, other than one resulting from an occupational injury, caused by the exposure to environmental factors associated with employment. Included are those acute and chronic illnesses or diseases that may be caused by inhalation, absorption, ingestion, or direct contact with toxic substances or harmful agents (U.S. DOL, 1990.)

All occupational illnesses must be recorded regardless of severity and if they result in the following:

- Death (must be recorded regardless of the length of time between the injury and death;
- One or more lost workdays;
- Restriction of work or motion;
- Loss of consciousness;
- Transfer to another job;
- Medical treatment other than first aid; or
- Termination of the injured or ill employee.

Fatalities

If an on-the-job accident occurs that results in the death of an employee or in the hospitalization of three or more employees, all employers are required to report the accident in detail to the nearest OSHA office within eight hours. (29 CFR 1904.8) The employer can file a report either orally or in writing and by telephone or telegraph. It must relate the circumstances of the accident, the number of fatalities, and the extent of any injuries. The Area Director may require additional reports concerning the accident.

Injury and Illness Records

"Employers must keep injury and illness records for each establishment. An establishment is defined as a 'single physical location where business is conducted or where services are performed.' An employer whose employees work in dispersed locations must keep records at the place where the employees report for work. In some situations, employees do not report to work at the same place each day. In that case, records must be kept at the place from which they are paid or at the base from which they operate" (U.S. DOL, 1991, p. 2).

Recordkeeping Forms

Only two forms are used for OSHA recordkeeping, form 200 and form 101. The OSHA No. 200 form, serves two purposes:

1) As the log of occupational injuries and illnesses on which the occurrence, extent, and outcome of cases are recorded during the year; and
2) As the summary of Occupational Injuries and Illnesses that is used to summarize the log at the end of the year to satisfy employer posting obligations.

A second form is the Supplementary Record of Occupational Injuries and Illnesses, or OSHA No. 101. It provides additional information on each of the cases that have been recorded on the log. (*See* Figures 3-1 and 3-2.)

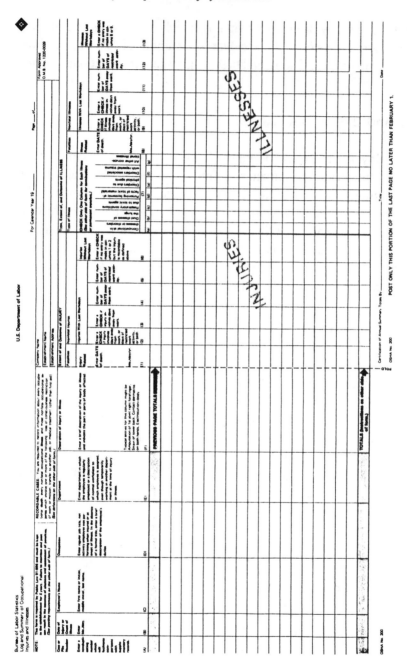

Figure 3-1. OSHA Form 200.

SUPPLEMENTARY RECORD OF OCCUPATIONAL INJURIES AND ILLNESSES

To supplement the Log and Summary of Occupational Injuries and Illnesses (OSHA No. 200), each establishment must maintain a record of each recordable occupational injury or illness. Worker's compensation, insurance, or other reports are acceptable as records if they contain all facts listed below or are supplemented to do so. If no suitable report is made for other purposes, this form (OSHA No. 101) may be used or the necessary facts can be listed on a separate plain sheet of paper. These records must also be available in the establishment without delay and at reasonable times for examination by representatives of the Department of Labor and the Department of Health and Human Services, and States accorded jurisdiction under the Act. The records must be maintained for a period of not less than five years following the end of the calendar year to which they relate.

Such records must contain at least the following facts:

1) *About the employer*—name, mail address, and location if different from mail address.

2) *About the injured or ill employee*—name, social security number, home address, age, sex, occupation, and department.

3) *About the accident or exposure to occupational illness*—place of accident or exposure, whether it was on employer's premises, what the employee was doing when injured, and how the accident occurred.

4) *About the occupational injury or illness*—description of the injury or illness, including part of body affected; name of the object or substance which directly injured the employee; and date of injury or diagnosis of illness.

5) *Other*—name and address of physician; if hospitalized, name and address of hospital; date of report; and name and position of person preparing the report.

SEE *DEFINITIONS* ON THE BACK OF OSHA FORM 200.

Bureau of Labor Statistics
Supplementary Record of
Occupational Injuries and Illnesses

U.S. Department of Labor

Form Approved
O.M.B. No. 1220-0029

Employer

1. Name

2. Mail address

3. Location, if different from mail address

Injured or Ill Employee

4. Name (First, middle, and last)   Social Security No.

5. Home address

6. Age   7. Sex: (Check one) Male ☐ Female ☐

8. Occupation

9. Department

The Accident or Exposure to Occupational Illness

10. Place of accident or exposure

11. Was place of accident or exposure on employer's premises? Yes ☐ No ☐

12. What was the employee doing when injured?

13. How did the accident occur?

Occupational Injury or Occupational Illness

14. Describe the injury or illness in detail and indicate the part of body affected.

15. Name the object or substance which directly injured the employee.

16. Date of injury or initial diagnosis of occupational illness

17. Did employee die? Yes ☐ No ☐

Other

18. Name and address of physician

19. If hospitalized, name and address of hospital

Date of report   Prepared by   Official position

OSHA No. 101 (Feb. 1981)

Figure 3-2. OSHA Form 101.

OSHA No. 200

"The OSHA 200 log is used for recording and classifying recordable occupational injuries and illnesses and for noting the extent and outcome of each case. The log shows when the occupational injury or illness occurred, to whom, what the injured or ill person's regular job was at the time of the injury or illness exposure, the department in which the person was employed, the kind of injury or illness, how much time was lost, and whether the case resulted in a fatality, etc" (U.S. DOL, 1990b, p. 2).

The *Code of Federal Regulations (CFR) Part 1904.2* provides the basic requirements for the OSHA No. 200:

A) Each employer, except those exempt, shall: (1) maintain in each establishment a log and summary of all recordable occupational injuries and illnesses for that establishment; and (2) enter each recordable injury and illness on the log and summary as early as practicable but no later than six working days after receiving information that a recordable injury or illness has occurred. For this purpose, form OSHA No. 200 or an equivalent which is as readable and comprehensible to a person not familiar with it shall be used. The log and summary shall be completed in the detail provided in the form and instructions on the back of form OSHA No. 200.

B) Any employer may maintain the log of occupational injuries and illnesses at a place other than the establishment or by means of dataprocessing equipment, or both, if

(1) There is available at the place where the log is maintained sufficient information to complete the log to date within six working days after receiving information that a recordable case has occurred as required by paragraph (A) of this section; and,

(2) At each of the employer's establishments, there is available a copy of the log which reflects separately the injury and illness experience of that establishment and current to a date within forty-five calendar days (U.S. DOL, 1990b).

Simply speaking, the employer has to maintain the log and write in all the recordable injuries and illnesses related to the work environment ("recordable" will be covered later). Enter the data from the incident as soon as possible, but at least within six working days after it occurred. OSHA estimates it will take four to thirty minutes to complete a line entry in the OSHA 200.

OSHA No. 101

For every injury or illness entered in the OSHA No. 200, it is necessary to record additional information on the supplementary record, OSHA No. 101. The supplementary record describes how the injury or illness exposure occurred, lists the objects or substances involved, and indicates the nature of the injury or illness and the part(s) of the body affected.

CFR 1904.4 provides the requirements for the supplementary record: In addition to the log of occupational injuries and illnesses provided for under section 1904.2, (OSHA No. 200) each employer shall have available for inspection at each establishment within six working days after receiving information that a recordable case has occurred, a supplementary record for each occupational injury or illness for that establishment. The record shall be completed in detail described in the instructions accompanying OSHA Form No. 101. Workers' compensation insurance or other reports are acceptable alternative records if they contain the information required by Form OSHA No. 101. If no acceptable alternative record is maintained for other purposes, Form OSHA No. 101 shall be used or the necessary information shall be otherwise maintained (U.S. DOL, 1990b, p. 17).

A copy of either the OSHA No. 200 or OSHA No. 101 can be obtained from the United States Department of Labor or the state department of labor-OSH Division. An employer may use other records in place of the 101 if those records contain the same information. Many companies depend on their workers' compensation forms to provide supplementary data.

Annual Survey

One other form that should be considered is the OSHA 200-S used by the Bureau of Labor Statistics (BLS) to keep track of occupational injuries and illnesses. Only selected companies receive copies of this questionnaire which provides information about the establishment's injuries and illnesses from the previous year. Employers will get the OSHA 200-S in the mail soon after the close of the year and will use the OSHA No. 200 as the source for completing this form. Companies of any size, including those with fewer than eleven employees may be required to complete these forms.

Recordability

The OSHAct provides a basic description of which cases are to be recorded. Recordkeeping regulations found in 29 CFR Part 1904 provide specific recording and reporting requirements which comprise the framework of the OSHA system. The regulations also expand upon the basic definition of recordability.

In a few situations, the criteria of the act, regulations, or the guidelines listed in this chapter may seem inappropriate. However, it would be virtually impossible to enact legislation, draft regulations, or issue guidelines that address every possible recordkeeping situation. Recordkeeping criteria must be sufficient to meet the needs of safety and health professionals maintaining complex programs while also remaining comprehensible to those maintaining records without the benefit of specialized safety and health training (such as some employers with small-sized establishments). There are approximately 75 million employees involved in the recordkeeping process through the posting and access provisions of the regulations.

Recordable Cases

Companies are required to record information about every occupational death, as well as

- every nonfatal occupational illness; and

- those nonfatal occupational injuries that involve one or more of the following: loss of consciousness, restriction of work or motion, transfer to another job, or medical treatment (other than first aid).

Some basic decisions must be considered before a case is to be documented and recorded. The decision-making process consists of five steps:

1. Determine whether a case occurred; that is, whether there was a death, illness, or an injury;
2. Establish that the case was work related; that it resulted from an event or exposure in the work environment;
3. Decide whether the case is an injury or an illness; and
4. If the case is an illness, record it and check the appropriate illness category on the log; or
5. If the case is an injury, decide if it is recordable based on a finding of medical treatment, loss of consciousness, restriction of work or motion, or transfer to another job.

Retention

The log and summary, OSHA No. 200, and OSHA No. 101, must be retained in each establishment for five calendar years following the end of the year to which they relate (U.S. DOL, 1990b).

Posting Requirements

"A copy of the totals and summary information for the year must be posted at each establishment wherever notices to employees are customarily posted. This copy must be posted no later than February 1, and kept in place until March 1. Even though there were no injuries or illnesses during the year, zero must be entered on the totals line, and the form posted." (U.S. DOL, 1990a, p. 8.)

Recordkeeping Variances

"Employers wishing to set up a recordkeeping system different from the one required by OSHA regulations may apply for a recordkeeping variance. Petitions must detail and justify the employer's intended procedures and must be submitted to the regional commissioner of BLS for the area in which the workplace is located. Similarly, in state-plan states, only BLS can grant a variance from recordkeeping requirements.

As with applications for variances from standards, an employer filing for a recordkeeping variance must give a copy of the application to the employees' authorized representative. The employer must also post a summary of the application wherever notices are normally posted. Employees have 10 working days to submit to BLS their own written data, views, or arguments." (U.S. DOL, 1990b, p. 8.)

In summary, remember that OSHA places special importance on posting and recordkeeping. The compliance officer will inspect the record of deaths, injuries, and illnesses which the employer is required to keep. The compliance officer will check to see that a copy of the totals from the last page of OSHA No. 200 has been posted and that the OSHA workplace poster (OSHA 2203) is prominently displayed. Where required, records of employee exposure to toxic substances and harmful physical agents are also examined for compliance with the recordkeeping requirements.

While the following items are not required for all OSHA standards, they should be recorded to accurately monitor and assess occupational hazards:

- Initial and periodic monitoring, including the date of measurement, for operations involving exposure; sampling and analytical methods used and evidence of their accuracy; number, duration, and results of samples taken; type of respiratory protective devices worn; and name, social security number, and the results of all employee exposure measurements. This record should be kept for 30 years.

- Employee physical medical examinations, including the name and social security number of the employee; physicians' written opinions; any employee medical complaints related to exposure to toxic substances; and information provided to the examining physician. These records should be maintained for the duration of employment plus 30 years.

- Employee Training. These records should be kept for one year beyond the last date of employment of that employee." (U.S. DOL, 1990c, p. 5.)

"Whoever knowingly makes any false statement, representation, or certification in any application, record, report, plan, or other document filed or required to be maintained pursuant to this Act shall, upon conviction, be punished by a fine of not more than $10,000, or by imprisonment, for not more than 6 months, or both." (U.S. DOL, 1990b, p. 28.)

CONCLUSION

Employers need to be careful in the application of laws in particular states. Although the types of awards vary, employers are subject to workers' compensation liability any time an employee is injured in the course of work. This could occur in a number of ways. In the case of an employee taking work home from the office, that employee is subjecting the employer to a possible claim while en route, as well as while working at home. Employees required to travel in the course of their jobs may also be covered by workers' compensation even during their off-hours.

Managers should make every effort to assure the safety of their employees at all times. When things go wrong, they must be prepared to track injuries and illnesses. These records must be kept in a timely manner and must include all required information. In most cases, the OSHA form 101 can be supplanted with the workers' compensation forms.

Questions

1. What is the value of records to the employer and to OSHA? Recordkeeping requirements by OSHA are heavily emphasized. Prior to a walkaround, in an inspection, the compliance officer will first look at records.

2. Is workers' compensation beneficial to the employer? Although it costs the employer money, why is it desirable from the employer's perspective?

3. How do complete company records provide protection to the employer? If they weren't required by law, should the employer still keep detailed occupational safety and health records?

References

Champa, Shirley A. Kentucky Workers' Compensation. Norcross, GA: The Harrison Company, 1982.

Kentucky Labor Cabinet. 1991. *Facts about occupational safety and health*. Frankfort, KY: Kentucky Labor Cabinet.

National Institute for Occupational Safety and Health. 1979. *Self-evaluation of occupational safety and health programs*. Washington, DC: NIOSH.

United States Department of Labor. 1990a. *OSHA compliance manual: What is OSHA*. Washington, DC: U.S. Government Printing Office.

United States Department of Labor. 1990b. *OSHA compliance manual: Recordkeeping guidelines*. Washington, DC: U.S. Government Printing Office.

United States Department of Labor. 1990c. *OSHA compliance manual: Inspections*. Washington, DC: U.S. Government Printing Office.

United States Department of Labor. 1991. *All about OSHA*. Washington, DC: U.S. Government Printing Office.

Chapter 4

Safety Related Business Laws

CHAPTER OBJECTIVES

After completing this chapter, you will be able to

- Understand the important legal terminology related to civil law.
- Identify the different types of torts.
- Understand product liability and associated risks.
- Understand product safety and the Product Safety Act.
- Identify the types of warranties.
- Understand contracts.

CASE STUDY

Larry Wilson's son was badly burned in a motorcycle accident. Larry decided to sue the manufacturer to see if he could be compensated for uninsured medical bills, pain and suffering on his son's behalf, and future expected care costs. The company settled out of court for three-quarters of a million dollars before the case ever went to trial. This is an interesting case example. In a criminal case, where an individual violates a law, a party is presumed innocent until proven guilty. In a product liability case, like this one, a party wins or loses based on the *preponderance of evidence*. Guilt is not necessarily involved. The party with the most convincing argument wins. It is not unusual for a case of this nature to be settled before trial if parties are willing and they can reach an agreement on terms. Additional legal expenses can then be saved. However, each is gambling that the company or individual will be financially better off with a settlement.

Incidentally, this case was to be tried before a jury, as many such cases are.

The really interesting and ironic parts of this case were the circumstances surrounding the accident. Wilson had purchased the motorcycle from a neighbor for fifteen dollars earlier that day. He wheeled the motorcycle over to his house, pushed it into the basement and closed the door. The next time he thought about the motorcycle was when he heard shouts coming from the basement, where he found his son lying on the floor, engulfed in flames.

A subsequent investigation by the defendant's attorneys found that the motorcycle gas tank had no cap when Wilson bought it; yet he stored it next to the water heater. He took no steps to warn his son, nor did he give the boy permission to play with the motorcycle. When his son tried to kick start the motorcycle, he dropped it. Defense attorneys speculated that gas spilled out at that time and was ignited by the nearby water heater. The boy was pinned by the weight of the cycle and suffered burns as a result. Additionally, no doors had been opened to permit the exhaust from the motorcycle to leave the house had it, in fact, started.

Most people, including the motorcycle manufacturer's legal counsel, believed that the manufacturer had no liability. Should anyone have been sued, it should have been the neighbor who sold the motorcycle without a gas cap. Before the case reached trial, however, the manufacturer decided to settle. The company managers and defense attorneys believed that once the case went to trial, a jury would award a large cash settlement to the Wilson family. Even though preponderance of evidence may not have favored the Wilsons, juries are often sympathetic to the plights of injured people regardless of fault. They perceive that manufacturers have unlimited funds and the injured people who are suing have little financial resources. This is the reason the manufacturer settled the case before it went to trial.

IMPORTANT TERMINOLOGY

Criminal law is addressed through the prosecution, and the penalty suffered is either a fine or imprisonment. Civil law is addressed in the courts through litigation and penalties include monetary damages. For example, John Smith is driving the company truck. When he hits the brakes, they fail to operate properly and the truck runs off the road and into a ravine. As a result of the accident, John suffers severe back pain and is unable to work for several months. John perceives that the benefits he receives are insufficient to compensate him for the inconvenience, pain, and suffering he has undergone. Under *civil* law, John contacts a local attorney, and after discussing it with her, decides to sue. John is seeking recovery beyond the workers' compensation benefits.

After meeting with his counsel, John sues his employer, as owner of the truck; the dealership where the truck was last serviced; the manufacturer of the truck; and the owner of the land where the truck went off the road. John is the *plaintiff* and each of the parties being sued are *defendants*. His lawyer advised John to follow the *deep pockets theory* and to sue everyone who might have money in the case. As in the Wilsons' case, John will only go after individuals who are most likely to have the money to pay. After *depositions* are given by each of the parties in the suit, the landowner is removed from the list because John's legal counsel did not believe that this party shared fault. Aside from the small piece of property, the landowner had few assets and the defense counsel believed that it was in the plaintiff's best interest to excuse him from the suit. The policy of some companies is to never settle out of court if they do not absolutely believe they are at fault. Their perception is that settlements encourage additional lawsuits. Other companies settle when possible to avoid the legal expenses associated with a long court case, as well as the adverse publicity that often accompanies a trial.

Lawsuits, like the ones mentioned above, are typically initiated following a tort. A *tort* is an act, or absence of an act, that causes a person to be injured, a reputation to be marred, or property to be damaged. These cases differ from a criminal case in that the purpose of a tort case is to obtain compensation for damage suffered. Criminal cases seek to punish the wrongdoer without compensation. A tort may give rise to criminal action or vice versa.

Today, any accident resulting in death, injury, or even property damage can bring about a law suit, criminal prosecution, or both. Anyone can sue another party anytime for practically any reason. Lawsuits range from the justified to the ridiculous. Most are a result of someone perceiving that they have been wronged and are seeking compensation.

From a safety perspective, one of the major concerns is liability. Liability is either voluntarily assumed, as by contract, or it is imposed as it would be following a lawsuit. Liability resulting from torts is of the second type. It is important to understand tort law because managers can sometimes unwittingly find themselves on the losing end of a lawsuit. The activities that can act as catalysts for lawsuits and that are difficult, if not impossible, to defend include the following:

Intentional torts include those where the defendants intended for their actions to cause the consequences of their act. Examples of intentional torts are assault, battery, false imprisonment, and defamation and are defined below.

Assault occurs when another person fears personal harm or offensive contact. If one employee threatens to hit another employee with a shovel, that employee may be guilty of assault. If a person is made to feel apprehensive or afraid that he will be hurt or offensively contacted, then the other party may be guilty of assault. A key element of apprehension is the knowledge that the assault occurred. If a worker shakes his fist at another worker or threatens another worker bodily in such a way that the other worker is unaware of the motions, then no assault has occurred.

Battery is harmful or offensive bodily contact. If an employee hits another employee or even grabs the employee by the shirt collar, that employee may be guilty of battery. The act may have caused little or no injury, but if it offends a reasonable person, then it may be considered battery.

False imprisonment occurs when a person is restrained within a fixed area against their will. If, however, the restrained individual has a means of exit available, then no false imprisonment has occurred. For example, merchants who restrain suspected shoplifters may be sued for false imprisonment. If, however, the merchant merely stands between the shoplifter and the door to discuss the issue, and the shoplifter has another means of exit, no imprisonment has occurred.

Defamation is the act of injuring another individual's reputation by disgracing or diminishing that person in the eyes of others. Defamation can occur through libel where the communication is written or broadcast through media such as radio or television. It can also occur through slander where the communication is spoken. Defamation is grounds for a lawsuit. The best defense against a defamation suit is the truth, but employees should be warned about the consequences of defamation lawsuits. A negative reference letter or an adverse report about someone given by telephone could trigger a defamation suit.

All of the above are examples of *intentional torts*. In order to successfully sue someone for an intentional tort, the plaintiff has to present a "preponderance of evidence" as opposed to proving the defendant guilty, as a prosecutor tries to do in a criminal case. In an intentional tort lawsuit, the verdict is awarded to the individual who presents enough evidence to convince the judge or jury that he should receive the verdict in his favor.

Individuals need to also be aware of *unintentional torts* when parties are sued due to their unintentional actions. Typically, the parties being sued are negligent. *Negligence* involves creating an unreasonable risk of harm, as opposed to intentional torts that deal with conduct that is likely to cause harm. The failure to exercise reasonable care under these circumstances can lead to a negligence suit.

Reasonable care is a relative term that is dependent on the circumstances and the party involved. The legal standard for reasonable care is the reasonable person who always acts prudently under the circumstances and is never careless or negligent. If the party happens to be a professional, then the standard is raised. For example, a person who is qualified to work in a profession or trade requiring special skills is expected to act in a manner consistent with that of a prudent and careful surgeon, carpenter, accountant, or safety professional. The skill level and expectations are higher for a professional than they are for the non-professional in a given area.

If standards of performance exist, then the professional should be aware of those standards and make every effort to comply with them. For example, accountants have certain standards which are maintained and published by the Financial Accounting Standards Board. Safety professionals have standards published by the Board of Certified Safety Professionals. When accountants, safety practitioners, or other professionals remain ignorant of the standards of their trades or choose to ignore those standards, they are increasing their risk of being sued and of being unsuccessful in the defense of a suit. Professionals are held to a higher standard of performance by the courts.

Even though a plaintiff can successfully present a case for negligence and a preponderance of evidence, a favorable award may not be received because of a successful defense by the defendant. Successful defenses that have prevented awards include contributory negligence, comparative negligence, and assumption of risk.

Contributory negligence occurs when the plaintiff helped cause or actually contributed to the loss. If an employee mops the floor in an area where people have the option to go around but choose not to, and a person slips, falls, and is injured on the wet floor, the person who fell contributed to the accident through negligence and may be denied an award.

In some states, the above award may simply be shared by the defendant and the plaintiff. Under *comparative negligence*, the award is apportioned based on the percentage of fault. Therefore, in a comparative negligence defense, if the defendant was found to be forty percent at fault and the plaintiff sixty percent at fault, and the award was for $100,000, the actual cash to be paid by the defendant would be $60,000. If the plaintiff's share of fault is found to be greater than fifty percent, then the plaintiff can receive no award.

PRODUCT LIABILITY

Within the last twenty-five years, the litigation costs surrounding product safety have evolved with some lawsuits reaching the million- or billion-dollar award level in damages. According to Seiden (1984, p.15), products should "prevent and/or mitigate personal injury and illness under reasonably foreseeable conditions of service, including reasonably foreseeable use and misuse." Much of the litigation evolved from the consumer movement of the late 1960s and early 1970s. One results of this movement was the passage of the Product Safety Act.

Product Safety Act

The Product Safety Act was officially enacted on October 27, 1972, with four main purposes:
1) To protect the public against unreasonable risks of injury associated with consumer products

2) To assist consumers in evaluating the comparative safety of consumer products
3) To develop uniform safety standards for consumer products and to minimize conflicting state and local regulations
4) To promote research and investigation into the causes and prevention of product-related deaths, illnesses, and injuries. (BNA, 1973, Appendix 3)

This Act is administered by the Consumer Product Safety Commission which promulgates consumer product safety standards. These standards are "reasonably necessary to prevent or reduce an unreasonable risk of injury associated with such product" (BNA, 1973, Appendix 7). The text of the Product Safety Act states that there are two requirements for product safety:

1) Requirements as to performance, composition, contents, design, construction, finish, or packaging of a consumer product
2) Requirements that a consumer product be marked with or accompanied by clear and adequate warnings and instructions, or requirements respecting the form of warnings and instructions (Colangelo et al., 1981, p. 1)

Companies manufacturing products that have not considered the above requirements place themselves at risk for product liability.

Product liability "describes an action, such as a lawsuit, in which an injured party (the plaintiff) seeks to recover damages for personal injury or loss of property from a seller or manufacturer (the defendant) when it is alleged that the injuries or economic loss resulted from a defective product." (Colangelo et al., 1981) Manufacturers, in turn, can effectively minimize product liability by developing products with safeguards against predictable types of defects, deficiencies, abuse or misuse. However, there are many unforeseeable factors that can hamper achievement of this goal. For example, injuries such as injuries to consumers using (or misusing) products can facilitate product liability litigation, thus increasing the manufacturer's responsibility for producing a safe product and the cost of that product to consumers.

Theories of Product Liability

There are four important theories that form the basis of product liability and provide a fundamental understanding of product liability litigation:

Strict liability is the concept that a manufacturer of a product is liable for injuries received due to product defects, without the necessity of a plaintiff to show negligence or fault (Hammer, 1972). In some instances, a party may be liable for damages caused even though there was no negligence or intentional tort. Some activities, even though they are desirable and often necessary, are so inherently dangerous that they are a risk to others regardless of how carefully they are conducted. In such cases, the purveyor of the activity is held to the concept of *strict liability* or liability without fault.

An example of strict liability is when a company stores explosives or even flammable liquids in large quantities, or emits noxious fumes. Once it is determined that the damages are linked to the activity, the manufacturer will generally be held liable regardless of circumstances.

Negligence is the failure of the manufacturer to exercise reasonable care. In addition, it also includes the responsibility to carry out a legal duty so that injury or property damage does not occur to another (Hammer, 1972).

Breach of express warranty occurs when the product does not meet the claims made by the manufacturer and, as a result, damage or injury occurs to another (Colangelo et al., 1981). When a company gives a statement of performance concerning a product, this statement becomes a part of the product. If the product fails to live up to the standards outlined by this statement, then the product breaches the *express warranty*. This warranty may be written, oral, or simply in the form of a demonstration. The assumption made by the buyer, and justifiably so, is that the product purchased will perform as well as was shown in the demonstration.

Breach of implied warranty occurs when a product fails to fulfill the purpose for which it was intended. An *implied warranty* is a warranty that is implied by the law rather than specifically made by the seller. The basis for this rests primarily on definitions found in the Uniform Commercial

Code. In essence, the seller may be held liable if the seller makes an improper recommendation regarding its use of the product (Colangelo et al., 1981). Even though no particular warranties are expressed, products are purchased for a particular purpose. When a product fails to fulfill that purpose, it is in violation of the implied warranty. If, for example, a consumer buys a balloon and the balloon will not hold air, it is in violation of the implied warranty. The assumption made by the purchaser is that all balloons hold air.

Manufacturers or sellers are not liable for all injuries that may result from a product. This would be the concept of *absolute liability*. However, these four basic theories clearly establish the product responsibilities of the manufacturer or seller in most states (Brauer, 1990).

Liabilities extend to the final user of the product, regardless of how many times it changes hands. The liability extends to wholesalers, retailers, and to the final user and is know as *vertical privity*. Injured bystanders would also be protected under the concept of *horizontal privity*. *Vertical priority* refers to parties up and down the distribution chain. *Horizontal priority* refers to users on a given level of the chain.

Lawsuits

In a product liability lawsuit, the plaintiff must present certain evidence to support the claim. With the exception of express warranty cases, the plaintiff must prove (Brauer, 1990):

1) The product was defective.
2) The defect existed at the time it left the defendant's hands.
3) The defect caused the injury or harm and was proximate to the injury.

In reference to the four theories of product liability, each has its own requirements for evidence presentation (Brauer, 1990):

- **Strict Liability**. No other evidence is required to establish a case.
- **Negligence**. The plaintiff must show that the defendant was negligent in some duty toward the plaintiff.

■ **Express Warranty and Implied Warranty**. The plaintiff must show that a product failed to meet the implied or express warranty or misrepresented claims for the product.

The manufacturer or seller, on the other hand, have a number of defenses at their disposal for the four theories (Brauer, 1990). They may try to show that although the product is dangerous, the danger itself is not a defect. The defendant could try to show that the plaintiff "altered the product or unreasonably misused it; or the product itself did not cause the injury." The defendant might also claim that "the product met the accepted standards of the government, industry, or self-imposed standards related to the product, to the claimed defects, and to the use of the product."

There are risks that manufacturers or sellers face when they put a product on the market. There are, however, a number of ways that liability can be minimized:

■ Defend the manufacturer in design, manufacturing, packaging, and the marketplace.

■ Assemble a team for product design review to thoroughly analyze the product for hazards and acceptable controls (Brauer, 1990).

■ Remove unreasonable dangers from products and environments, and prevent defects from reaching the marketplace, and account for the use and misuse of the product and consider hazards, potential injury, compliance with standards, and claims for products.

■ Ensure that warnings identify remaining hazards, and provide instructions necessary for user protection.

At every step in the product life cycle, beginning with product conception, the safety professional should have input. (See Chapter 9 for additional information.) Safety should not be an afterthought in the design and manufacture of products if liability is to be minimized. Safety must be built into each and every product that leaves the facility.

CONTRACTS

A *contract* is a legally enforceable promise. It may be expressed, implied in fact, or quasi-contractual. Be aware that contracts can be made in a number of ways. *Express* contracts are made when a person makes an oral or written commitment to another to enter into a binding relationship. *Implied in fact* contracts occur when parties indicate through their actions, rather than words that they intend to agree. *Quasi contracts* occur when one party would be enriched unjustly if permitted to enjoy something received from another party. The party may have to make reasonable restitution.

In order for a contract to be valid, four conditions must be present. First of all, there must be an agreement; this requires an *offer*. An offer states, in reasonably certain terms, what is promised and what each party will do. The intention of the offering party is considered by the courts. For example, if the person is joking, asking for details, or simply negotiating, an offer may not take place. On the other hand, a proposal may be an offer even though all of the terms have not been specified.

Once an offer has been extended, *acceptance* may be made by the other party, or the other party may make a counter offer. In any case, in order for the contract to come into existence, there must be an agreement to the terms of the offer and then acceptance. If the offer specifies that it must be accepted in a particular way in terms of time or means, unless the terms are adhered to, there is no contract.

Typically, there is no enforcement of the contract unless there is *consideration*. Consideration is exchanged in terms of goods, services, or restraint from doing something. Both parties must be obliged to provide something but the value provided by each party does not have to be equal. Sometimes a promise is exchanged for a small sum, but if fraud or improper conduct is evidenced, the contract is unenforceable.

All parties to the contract must have the legal *capacity* to enter into the contract. They must be of legal age, sober, and of sound mind. If, for example, a party enters into a contract with a minor, one who is typically under the age of 18, the contract may be *void*. There are exceptions to the above conditions but these are generally upheld. If the contract has an illegal intent, or calls for some action that violates public policy, then it is typically void.

Enforcement of contracts occurs under civil law. When one party fails to live up to her part of the contract, the other party may attempt to enforce the terms of the contract by bringing a lawsuit. The courts may require an award of money or may force a party in the contract to perform the promise specified. This remedy, known as *specific performance*, may be necessary when a monetary remedy would be insufficient to compensate for the unique requirements of the contract. For instance, if a specific piece of real estate or a rare object of art was involved, the court may require that particular property to be delivered.

Certain points of law are sometimes pitfalls for the unaware. For example, insurance contracts and some leases are considered *contracts of adhesion*. These documents are often offered to the purchaser or lessee. Because greater power rests with the author of the document, ambiguities and language that is subject to interpretation will typically be construed against the writer. In such cases, the contract writer is considered to be the expert, so interpretations will tend to favor the weaker party such as the insured or the lessee.

Exculpatory or *hold harmless* clauses are sometimes written to limit or eliminate liability in certain contracts. For example, when a company hires a contractor to perform work in its business, the contractor may require that the company first sign an exculpatory agreement that states that the contractor will not be responsible for any damages the contractor causes. Clauses in contracts similar to this do not provide much protection for contractors or others whose intentional acts or negligence cause injury or damage. Companies still use them in the hope that they may deter another party from actually suing them.

Safety professionals should be aware of the *parol evidence* rule which states that the written contract supersedes any oral promises that were made. When the written contract is agreed to, it will govern the contract. For example, a salesperson's promises, if not included in the contract, will usually not be enforced. It is, therefore, worthwhile to review all written contracts that have been based on a sales pitch. Insurance contracts, leases, and service or maintenance contracts are typical of the types of documents that might differ from the buyer's expectations.

INSURANCE

Companies can build safety into a product every step of the way. No unusual warranties may be given, and the sales people may do an excellent and thorough job of explaining the product features. Buyers can even sign an exculpatory clause. However, even with the best products and the best of intentions, companies will still be sued. Sometimes, the only thing that stands between them and potential bankruptcy is insurance. Insurance companies not only pay up to specified amounts when lawsuits are lost, but they also provide defense attorneys for the defendant. It is sometimes less expensive to defend against a lawsuit than to simply pay up. Insurance is often the best protection against spurious lawsuits.

CONCLUSION

The safety practitioner needs to be aware of the basic legal environment surrounding the workplace. Although they often depend on legal counsel for specific advice, an awareness of the laws likely to affect the profits of the company is critical. Knowledge of potential pitfalls and methods to avoid those traps can only help to increase company profitability.

The concepts discussed in this chapter are complex and hold many exceptions and potential problems for manufacturers and consumers. Complete details are beyond the scope of this text. Additional research or legal consultation is recommended for answers to specific problems or questions.

Questions

1. Is the job of the safety manager one that requires concern for contracts, liability, and insurance, or should the safety practitioner simply be concerned with "safety?"

2. Do you think that there should be a cap on the amounts of jury awards which may be given beyond actual losses? Why or why not? Who pays for these awards?

3. How does a concern for torts become a part of the safety practitioner's concerns? Are you aware of situations where an employee or an agent of a company has committed a tort against another and the company was held financially responsible? Who pays in this case?

4. Do you ever use exculpatory or hold harmless clauses in your work? Are you ever asked to sign them when you rent, purchase, or use property? What are the circumstances when you are likely to see them?

References

Brauer, Roger L. 1990. *Safety and health for engineers*. New York, NY: Van Nostrand Reinhold.

Bureau of National Affairs. 1973. *The consumer product safety act: Text, analysis, legislative history*. Washington, DC: Bureau of National Affairs.

Colangelo, V. J. and Thornton, P. A. 1981. *Engineering aspects of product liability*. Metals Park, OH: American Society for Metals.

Hammer, W. 1972. *Handbook of system and product safety*. Englewood Cliffs, NJ: Prentice-Hall.

Seiden, R. M. 1984. *Product safety engineering for managers: A practical handbook and guide*. Englewood Cliffs, NJ: Prentice-Hall.

Chapter 5

Accident Causation and Investigation: Theory and Application

CHAPTER OBJECTIVES

After completing this chapter, you will be able to

- Explain the benefits of understanding Accident Causation Theory.
- Define the terminology associated with Accident Causation Theory.
- Identify the activities involved in Risk Assessment.
- Compare and contrast the various Accident Causation Theories.
- Explain the purpose of Accident Investigation.
- List the activities involved in Accident Investigation.

INTRODUCTION

Safety and health professionals, or individuals in an organization that have been given that assignment, are responsible for identifying, evaluating, and controlling hazards in the workplace. To be effective, this member of management must determine the risk associated with hazards found at the facility. The professional then sets into motion the necessary activities required to eliminate those hazards before losses occur.

One way that safety and health professionals prepare for this proactive approach to accident prevention is by becoming familiar with various theories that examine why accidents occur. Increasing awareness of accident causation prepares the professional to be more vigilant to the human variables or workplace factors that can result in accidents. In addition, knowledge and awareness of these theories permit the professional to more thoroughly recognize and communicate organizational safety problems. Individuals with little safety and health experience might pass by a punch press leaking lubricating oil on the shop floor and recognize a slipping hazard. For the more experienced professional, however, these clues would raise several questions. What is the lubricant? Is it a fire hazard? Is it a health hazard? What do the operators use to clean the spillage? Is this solvent a health hazard? Are hazard communication programs, training and personal protective equipment required as a result of these products? Should we consider different lubricants and solvents? What about the punch press? Is it guarded? Are there repetitive motion or other ergonomic hazards associated with the operation of the press?

It is possible to greatly expand the list of questions that the safety and health professional might wish to ask, given the few signs provided in this simple scenario. The point of this hypothetical exercise is to show that the more information and experience that individuals' possess, the more safety and health hazards that they will anticipate or recognize. Accident causation theories can provide the tools to improve your competency as a safety and health hazard "Sherlock Holmes."

There are many different theories associated with the cause of accidents. These theories range from very simple to very complex. Some of these theories focus on the employee and how their action or lack of action contributes to accidents. Other theories focus upon management and their responsibilities for preventing conditions that will lead to accidents. Remember that theories are not facts. Theories must be viewed as tools that help predict relationships that may exist in the future. Of primary importance, once a review of these various and wide ranging theories has been completed, is the realization that accidents are not events that happen by chance. Accidents have specific causes. By applying one or more of these theories, the safety and health professional is in a better position to predict the possibility of accidents in the organization and initiate activities necessary to prevent their occurrence.

It must be noted that not everyone understands the relationship that exists between the causes and effects of accidents. Some members of an organization may fail to appreciate the dangers of ignoring workplace hazards and accidents. These individuals may view accidents as part of the cost of doing business. For these individuals, increased production will compensate for these *"uncontrollable problems."*

This short-sightedness is unfortunate in light of the spiraling cost of health care, liability resulting from negligence, citations levied by OSHA, or costs associated with damaged raw and finished products. Ignoring safety requires added production in order to make up for the losses that will be experienced (refer to the chapter on Safety Management for more information on this subject). This economic perspective does not take into consideration the human suffering associated with this type of decision.

For many individuals in the safety and health profession, *accidents represent a failure in the system or a management problem in the organization*. There are important reasons to take this approach. First of all, consider that management "holds all the cards." Management advertises for, hires, places, trains, and supervises the workers. If the worker makes a mistake, both OSHA and the courts will typically hold management responsible. In an accident investigation, management may blame the worker. This is referred to as the "pilot error syndrome;" that is, "Blame the pilot because he is dead and cannot defend himself." Once the worker or pilot is blamed, management and the management system is off the hook. Therefore, there is no reason to continue to search for why the accident occurred. It is important to emphasize that the safety practitioner is not looking for a place to assign blame, but for errors in the system. These errors can be addressed so that problems and accidents will be prevented in the future.

If organizational management does accept this concept, the safety and health professional can initiate a team approach to vigorously eliminate workplace hazards. As a result, the frequency and severity of injuries and losses will also be reduced. When the frequency and severity of injuries and losses are reduced, inefficiency and waste are reduced. All of these factors ultimately impact on the company's bottom line--profits.

THE CONCEPT OF RISK AND ACCIDENTS

As first defined in chapter one, *risk* is a term that describes both the probability and severity of a loss event. *Probability* refers to the likelihood of an event taking place. When used as part of a risk assessment tool in system safety, probabilities can be categorized as Frequent, Probable, Occasional, Remote, and Improbable (Roland and Moriarity, 1990). By studying safety related data, it is possible to determine statistical trends for a variety of factors such as the types of accidents (falls to the same level, struck by falling object, or overexertion injuries to identify just a few examples), or location where injuries are taking place (warehouse, assembly, paint shop, etc.). Using the data on the frequency of occurrence of these events, it is possible to classify injuries, property damage, or other loss factors in terms of one of the five probability categories.

Risk = Probability x Severity

It is important to point out that probability and frequency of loss events are just half of the risk picture. The severity of the loss event must also be considered. *Severity* refers to the magnitude of the loss in a given period of time. When used as part of the risk assessment tool in system safety, the severity of a particular condition is classified as one of four categories: Catastrophic, Critical, Marginal, or Negligible. These four categories correspond to death or loss of a system, severe injury or major damage, minor injury or system damage, and no injury or system damage respectively (Roland and Moriarity, 1990). When viewed on a severity continuum, near misses, scratches not requiring first aid or brief assembly line stoppage might be considered negligible reactions to an incident. At the opposite end of the severity continuum, multiple fatalities or an explosion that destroys an entire building represent catastrophic loss events.

Bear in mind that all companies face risk and the resultant potential losses on a daily basis. They willingly accept them, even welcome them with the hopes of gaining financial return. All new marketing and business ventures are examples of businesses taking risks. These are considered *speculative loss exposures* because they offer the opportunity for gain, as well as for loss. Exposures that offer potential for loss with no opportunity for gain are referred to as *pure loss exposures*. Examples of pure loss exposures would include exposures to theft, fire, or accident.

Determining what exposures a company faces is a formidable task in itself, but it needs to be undertaken. Companies may choose to bring in an outside consultant or they may depend on the advice of a representative such as a *loss control* expert from their insurance company. This person will review their activities, procedures, and processes to help determine where exposures are occurring. Once these have been identified, the company must decide which ones are critical and what priorities are important to the company. Even large corporations have limited resources, and they must make decisions as to where to commit those resources. While controlling losses may seem to the safety manager like the most important way to utilize resources, the marketing director may feel that a new product introduction should take precedence.

Establishing priorities will allow the safety professional to determine which problems to attack first and which ones are worth spending more money on. Non-first aid injuries that occur with great frequency would appear, at first glance, to be events that do not rank very high on the safety professional's "things-to-do" list. On the other hand, confined space entry fatalities that are occasional in occurrence would be of great concern because employee deaths are often the result observed in those incidents. Multiplying the *probability* of loss from a fire or any event in a given year's time and the likely amount to be lost from such as fire or event (severity) gives the safety practitioner a relative gauge as to where to place resources for controlling losses.

By considering both probability and severity, the risk assessment provides the safety practitioner with a much sounder perspective from which to judge the significance of hazards. Once the risk assessment is performed, it is then possible to determine the types of controls that would most effectively eliminate the hazards.

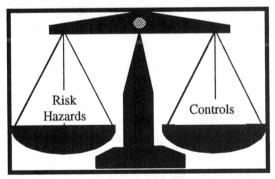

Figure 5-1. An illustration of the balance between risks and hazards in the workplace and controls necessary to minimize their efforts.

There are a number of control techniques available for treating loss exposures. These can be broken into two categories: loss control techniques and financing techniques. Loss control techniques include the approaches mentioned in Chapter 6, Introduction to Industrial Hygiene. A company may choose engineering, administrative, or personal protective equipment to deal with losses. Engineering controls such as ventilation systems to reduce explosive vapor levels would be one option available to the professional for controlling hazards. Administrative controls, limiting exposure to toxic materials, would be a second method for the protection of employees. Personal protective equipment (PPE) is the last line of defense against hazards in the work place. Refer to Chapter 6 for a detailed examination of these three types of control methods.

A company might try to *avoid* the loss altogether. If a company is in the business of producing football helmets, it may choose not to market the product due to the potential liability it would face if a wearer of one of its helmets was injured. Instead, it may choose to produce novelty lamps that look like football helmets to limit its liability exposure.

Sometimes a company can reduce exposure by *substitution*. Instead of using a strong acid as a solvent to remove excess glue from a finished product, a company may choose to use citric acid to remove glue which reduces the likelihood of a worker being injured.

Occasionally, the company may choose to *transfer* the liability to another party, rather than run the risk of loss itself. If removing glue could not be accomplished safely in the plant, the company may choose to have the product shipped to a contractor who would remove the glue in the contractor's plant. If the contractor's workers are overcome by vapors from the solvent, then the contractor would typically hold the liability.

Another form of transfer is *insurance*. Insurance is designed to permit the company to shift the financial consequences of the risk to an insurance company. By paying the insurance company's *premiums*, the organization can expect specified *benefits* in the event of a loss. The insurance company enters into similar relationships with a number of companies so that it can accurately estimate its own losses. When an insured company turns in a *claim*, the insurance company is able to pay because it receives enough money from similar companies who have not had losses to cover its claims and still make a profit.

Often companies simply retain their loss exposures without dealing with them. This may be a result of ignorance or choice. When companies choose to retain their own exposures, they may ignore them, or attempt to reduce them using one of the methods already mentioned, or they may, in fact, *self-insure*. Ignoring the risks may make the owners more confident, but dealing with the risks will make them more prepared for loss. Self-insurance is simply no insurance; the company retains the loss exposure. It should only be undertaken by companies with the financial resources necessary to absorb potential losses.

All of the above assume that the accident will occur. The safety practitioner can use accident causation models to pinpoint hazards in the occupational environment. Systematic, proactive hazard identification will assist the practitioner in establishing loss control strategies and determining the cost-benefits of the controls to be implemented.

The following review briefly examines some of the most popular accident causation theories. References found at the end of this chapter provide the student with resources to examine this area in greater detail. The intent of this review, however, is to provide the reader with examples of accident causation theories representative of past and current thinking on the subject. The "bottom line" benefit of this review is to provide the reader with the tools necessary to seek out and eliminate the causes of accidents.

ACCIDENT CAUSATION THEORIES

Single Factor Theory

The Single Factor Theory states that there is a single and relatively simple cause for all accidents. A good example of using this theory would be in determining the cause of worker hand lacerations. Because utility knives are used in the operation, the cause of these accidents would not necessarily stop the problem. Other contributing factors such as the product or the work methods, as well as corresponding corrective actions, are overlooked making this theory virtually useless for accident and loss prevention. Unfortunately, the Single Factor Theory often represents the thinking of some in management, and it is, therefore, important for the safety professional to point out the limitations of this theory to those members of management.

Domino Theories

There are several domino theories of accident causation. While each domino theory presents a different explanation for the cause of accidents, they all have one thing in common. All domino theories are divided into three phases: (1) the pre-contact phase, (2) the contact phase, and (3) the post-contact phase. The pre-contact phase of all domino theories refers to those events or conditions that lead up to the accident. During the contact phase the individual, machinery, or facility comes into "contact" with the energy forms or forces which are beyond their physical capability. The post-contact phase refers to the results of the "accident" or energy exposure. Physical injury, illness, production downtime, damage to equipment and/or facility, and loss of reputation are just some of the possible results that can occur during the post-contact phase of the domino theory.

Domino theories represent accidents as predictable chronological sequences of events or causal factors. Each causal factor builds upon and affects the others. If allowed to exist without any form of intervention, these hazards will then interact to produce the accident. In dominos games where they are lined up and the first one is knocked over, it sets into motion a chain reaction of events resulting in the toppling of the remaining dominos. Just as dominos that are lined up and standing on end can be toppled by knocking over the first domino in the line, accidents, according to the domino theories, will result if the sequence of pre-contact phase causes exist.

Heinrich's Domino Theory

The original domino theory of accident causation was developed in the late 1920s by H. W. Heinrich. Although written over seventy years ago, his work in accident causation is still the basis for several contemporary theories.

According to Heinrich's early theory, the following five factors influence all accidents and are represented by individual dominos:

1. Negative character traits that might lead people to behave in an unsafe manner can be inherited or acquired as a result of the social environment.
2. Negative character traits are why people behave in an unsafe manner and why hazardous conditions exist.
3. Unsafe acts committed by people and mechanical or physical hazards are the direct causes of accidents.
4. Accidents that result in injury are typically caused by falls and the impact of moving objects.
5. Typical injuries resulting from accidents include lacerations and fractures.

The two key points in the Heinrich's domino theory are: (1) injuries are caused by the action of preceding factors, and (2) removal of the events leading up to the incident, especially employee unsafe acts or hazardous workplace conditions, prevents accidents and injuries. Heinrich believed that unsafe acts caused more accidents than unsafe conditions. Therefore, his philosophy of accident prevention focused upon eliminating unsafe acts and the people-related factors that lead up to injuries (Brauer, 1990).

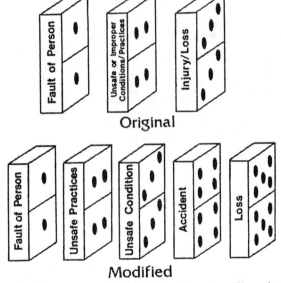

Figure 5-2. An illustration of Heinrich's Domino Theories of Accidental Causation.

Bird and Loftus' Domino Theory

Bird and Loftus (1976) updated the domino sequence to reflect the management's relationship associated with the causes and effects of all incidents. Bird and Loftus' theory uses five dominos that represent the following events involved in all incidents:

1) Lack of Control - Management.

An event could be caused by the lack of "control" by management. Control in this instance refers to the four functions of a manager: Planning, organizing, leading, and controlling. Purchasing substandard equipment or tools, not providing adequate training, or failing to install adequate engineering controls are just a few examples represented by this domino.

2) Basic Cause(s) - Origin(s).

The basic causes are frequently classified into two groups: 1) personal factors such as lack of knowledge or skill, improper motivation, and physical or mental problems, and 2) job factors including inadequate work standards, inadequate design or maintenance, normal tool or equipment wear and tear, and/or abnormal tool usage such as lifting more weight than the rated capacity of an overhead crane. These basic causes explain why people engage in substandard practices.

3) Immediate Causes(s) - Symptoms.

The primary symptoms of all incidents are unsafe acts and unsafe conditions. "When the basic causes of incidents that could downgrade a business operation exist, they provide the opportunity for the occurrence of substandard practices and conditions (sometimes called errors) that could cause this domino to fall and lead directly to loss." (Bird and Loftus, 1976, p. 44).

4) Incident - Contact

"An undesired event that could or does make contact with a source of energy above the threshold limit of body or structure." (Bird and Loftus, 1976). The categories of contact incident events are often represented by the 11 accident types. The eleven accident types include: struck-by, struck-against, contact-by, contact-with, caught-in, caught-on, caught between, foot-level-fall, fall-to-below, overexertion and exposure (ANSI Z 16.2).

5) People - Property - Loss
Loss is the adverse results of the accident. It is often evaluated in terms of property damage, as well as the effects upon humans such as injuries and the environment.

The central point in this theory is that management is responsible for the safety and health of the employees. Like Heinrich's theory, the Bird and Loftus domino theory emphasizes that the contact incidents can be avoided if unsafe acts and conditions are prevented. Examining the operation of an organization, using the first three dominos will identify the conditions that permit the opportunity for incidents to take place. By focusing on the work place and ensuring that the appropriate management activities are performed, accidents and related losses can be eliminated according to this theory.

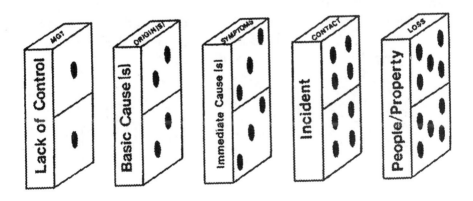

Figure 5-3. An illustration of Bird's Theory of Accidental Causation.

Marcum's Domino Theory

According to C. E. Marcum's 1978 *Seven Domino Sequence of Misactsidents,* a Mis-actsident was an identifiable sequence of misacts associated with *inadequate task preparation* which could lead to *substandard performance* and *miscompensated risks.* This permitted individuals and facilities to come in contact with harmful agents, energy forms, forces, or substances in ways that initiated adverse reactions sufficiently extensive that unwarranted losses were sustained and resultant costs incurred.

Like the Bird and Loftus theory of accident causation, Marcum focused upon management responsibility for protecting employee safety as well as preventing the downgrading of an organization. Downgrading of an organization included losses such as equipment and facility damage as well as the less obvious factors such as loss of reputation. Losses that result from harmful contact incidents were examined in more detail. For example, this theory attempted to examine management emergency response protocols to ensure that sustained losses and costs were minimized. Marcum kept these two post-contact-phase components separate to permit the closer monitoring of these two variables during accident analysis activities. Throughout this theory, Marcum focused upon the human element of "miss-acts." This includes misacts of the employees who failed to recognize or appreciate risks in the workplace, as well as misacts of organization management who permitted those risks to go unrecognized, unappreciated, and/or uncorrected.

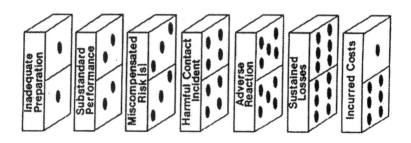

Figure 5-4. An illustration of Marcum's Theory of Accidental Causation.

It is important to point out that Marcum used the term Misactsidents to emphasize the deterministic aspects of his accident causation theory. Accidents, Marcum believed, were considered by most people to be events that occurred by chance. These were events that the average person on the street thought he had no control over or considered them as "acts of God." Marcum emphasized that accidents had specific causes and could be controlled. He emphasized educating employees on the concept that safety could and must be managed.

Multiple Causation Accident Theories

Multiple Factors Theory

Grose's multiple factors theory uses four Ms to represent the factors that cause an accident: machine, media, man, and management (Brauer, 1990). Machine refers to tools, equipment, or vehicles that may contribute to the cause of an accident. Media takes into account the environmental conditions surrounding an accident such as the weather condition or walking surface location. Man deals with the people and human factors that could contribute to the incident. Management takes into account the other three Ms looking at the method used to select equipment, train personnel, or ensure a relatively hazard-free environment.

The multiple factor theory examines characteristics surrounding the four Ms. Examination of machinery characteristics include the design, shape, size, or specific type of energy used to operate the equipment. Characteristics of man are psychological state; gender; age; physiological variables including height, weight, or condition; and cognitive attributes such as memory, recall, or knowledge level. Snow or water on a roadway, temperature of a building, and outdoor temperature outside can be characteristics of media. Characteristics of management could be safety rules, organizational structure, or policy and procedures.

Multiple factors theories attempt to identify specific workplace characteristics that can reveal underlying, and often hidden, causes of an accident. They seek to point out the hazardous conditions that exist in an operation. When viewed as a whole, these characteristics direct the investigator's attention to the specific causes of an accident.

Systems Theory of Causation

One variation of the Multiple Causation Theory is R.J. Firenzie's theory of accident causation. Firenzie's theory is based on the interaction between three components: person, machine, and environment. How these three components interface with each other determines the likelihood of an accident occurring.

In addition, information, decisions, and perception of risks have a bearing on the person performing a job. These human variables will combine with the machine hazards and the environmental factors affecting the likelihood of an accident.

For example, as a person operates a noisy bulldozer on a hot day, other activities must take place for the operator to safely and effectively perform the job. The person consciously or unconsciously may collect information, weigh the risks, and make a decision as to how to perform the task in a specific way. How close should the dozer get to the twenty-foot-high "spoil bank" or the high-energy electrical power lines? How fast should it be moving? Countless numbers of decisions are made by the operator based upon the operator's knowledge and experience. Every time this process occurs, there is a likelihood that an accident will occur.

Psychological/Behavioral Accident Causation Theories

Goals Freedom Alertness Theory

Dr. Willard Kerr's theory of accident causation regarded an accident as a low-quality work behavior. He considers it to be similar to production waste during manufacturing, only the scrap happened to be human. Raising the level of quality and safety involves raising the level of worker awareness. According to Kerr, alertness can only be obtained within a positive organizational culture and psychological climate. The more positive the workplace climate, the greater the level of alertness and work quality. At the same time as this state of alertness decreases, there is an increased probability of an accident.

Motivation Reward Satisfaction Model

This theory of accident causation builds upon Dr. Willard Kerr's Goals Freedom Alertness theory. Stated simply, the "freedom to set reasonably attainable goals is typically accompanied by higher-quality work performance."(Heinrich et al., 1980, p. 44) If an accident occurs, according to this theory, it is due to a lull in alertness. Combining this with the belief that the safety performance of an employee depends on the employee's degree of motivation and capability to work, Dr. Petersen concludes that factors which affect these variables will either promote or prevent accidents. Because motivation is influenced by multiple variables, then performance, including job safety, is susceptible to shifts in success.

According to this theory, rewards are the factor that has the greatest effect upon performance. Rewards can originate from multiple sources and can be physical and/or psychological. Money or praise are not considered to be the primary motivation factors. Rewards including doing a good job, learning new skills, expanding personal knowledge, and being a member of a successful team are just a few of the numerous intrinsic reinforcers associated with enriched jobs. Therefore, if employees see the rewards that they obtain from their work as equitable, then there is an increased likelihood of motivation that will produce positive safety results.

Human Factors Theory

The Human Factors Theory is based on the concept that accidents are the result of human error. Factors that cause human error are

1. Overload
2. Inappropriate activities
3. Inappropriate response

Overload occurs when a person is burdened with excessive tasks or responsibilities. For example, not only must the employee perform her job, but she must also handle excessive noise, stress, personal problems and unclear instructions.

Inappropriate activities is another term for human error. When individuals undertake a task that they are not properly trained to do, they are acting inappropriately.

Inappropriate response is when, for example, an employee detects a hazardous condition but does not correct it, or when an employee removes the safeguard from the machine to increase productivity. Overload, inappropriate activities, and inappropriate responses are all human factors that cause human error and, ultimately, accidents.

Energy-Related Accident Causation Theories

Energy Release Theory

This theory treats accidents as a physical engineering problem. Accidents result when energy that is out of control puts more stress on a person or property than they can tolerate without damage. Accidents can be prevented by controlling the energy involved or changing the structures that the energy could damage. Haddon's ten strategies were expanded to twelve by William Johnson when they are applied to the Accident Investigation strategy referred to as Management Oversight and Risk Tree (MORT). Johnson's barriers to accident causing energy include:

1) Limit the Energy. Example: Limit the amount of flammable or combustible materials that are stored on the shop floor. Use low-voltage equipment.
2) Substitute a safer energy form. Example: Use nonasbestos brake pads. Select non-flammable or nontoxic solvents.
3) Prevent the build-up. Example: Fuses and Circuit breakers. Gas detectors.
4) Prevent the release of the energy. Example: Toe boards on scaffolds to prevent tools from striking people or objects below.
5) Provide for slow release. Example: Safety-release valves
6) Channel the release away; separate it in time or space. Example: Electrical grounding, demarcating Inorganic Arsenic regulated areas.
7) Place a barrier on the energy source. Example: Machine guards, acoustic enclosures.
8) Place a barrier between the energy source and the persons or objects to be protected. Example: Rails on elevated surfaces; Fire doors.
9) Place a barrier on the persons or objects to be protected. Example: Personal Protective Equipment (PPE) and respirators.
10) Raise the injury or damage threshold. Example: Acclimatize to hot or cold work environment.
11) Ameliorate the effects. Example: Administrative controls such as job rotation to
   reduce the duration of exposure to loud noise.
12) Rehabilitate. Example: Treat injured employees or repair objects that were damaged.

Haddon and Johnson focus on energy as the source of the hazard. By identifying the energy sources and preventing or minimizing the exposure, accidents will be prevented.

ACCIDENT INVESTIGATION

The purpose of any accident investigation is to determine the causes that resulted in the accident. This information should then be used for the prevention of similar accidents. Accident investigation is concerned with fact finding, not fault finding.

During the accident investigation, it is important to find out answers to the questions: *who, what, where, when, how,* and *why.*

Who questions include:
- Who are the victims?
- Who are the witnesses?
- Who has any information that will help determine the actual causes of the incident?

What questions include:
- What events lead up to the accident?
- What were victims and witnesses doing prior to and during the incident?
- What did individuals notice that may have a bearing on the incident?
- What were the backgrounds and experiences of all the parties involved?

Where questions include:
- Where were victims and witnesses prior to and during the incident?
- Where was equipment and/or machinery?
- Where was the PPE or the locks and tags for energy sources?

When questions include:
- When did the incident occur?

- When did you notice important elements associated with the incident? When did you become concerned that a problem existed?

How questions include:
- How did the incident take place?
- How did the victims and witnesses react in given situations?
- How did you first learn of the incident?

Why questions include:
- Why, in your opinion, did the incident take place?
- Why were particular methods used to perform a task?
- Why were conditions existing at the time of the incident?

When an investigation team first arrives at the scene, the key elements must be searched out for possible clues. Those key elements have been identified as: personnel, tools/equipment, raw materials or finished product, and structure and environment. All of these elements can provide evidence that will permit the accident investigation team to gather evidence that will assist in the investigation.

Preplanning and preparation is of vital importance. Knowing the roles and responsibilities during an investigation is crucial. In addition, preparation that permits prompt arrival at the accident scene, so that evidence does not disappear or witness recollections are not lost over time, is imperative. Ensure that all of the tools and equipment necessary to conduct the investigation are organized and available at any time of the day or night. Confirm that all of the appropriate individuals are contacted from the plant manager or corporate CEO to OSHA and EPA officials. Provide training and mock drills to give the knowledge and practice vital for effective responses to emergencies. These are many of the preparatory activities that are required to increase the effectiveness of saving lives and protecting the evidence necessary to determine causes and prevent recurrences.

The following is a brief protocol of some of the steps that can be followed at a facility to prepare for accident investigations.

When an accident occurs:

1. Notify the following individuals immediately (*list important individuals here*)

2. Secure the scene to prevent additional accidents; bring in accident trained technical personnel to determine:
   - ☐ damage to and hazard from power, gas, and fluids distribution systems,
   - ☐ the structural integrity of the building and equipment
   - ☐ the best way to remove or make harmless explosives and/or hazardous materials

   caution: make sure the position of switches, equipment, and materials are recorded before they are moved or removed

3. Evaluate the condition of any injured personnel; determine
   - ☐ what is the degree of injury
   - ☐ what must done immediately to save life
   - ☐ what should be done to relieve suffering until the injured individual can be removed to a medical facility
   - ☐ what can be done to remove all danger of increased injury

   caution: move an injured person only if there is danger of further injury

4. Identify the elements at the accident scene
   - ☐ people involved
     - injured
     - principals
     - witnesses
   - ☐ equipment involved
     - in use

- standby
- secured or standing
- materials involved
- in use
- ready for use
- stored in area
☐ environmental factors
  - weather
  - lighting
  - heat
  - noise
☐ any additional contributing factors
☐ keep detailed notes for reference

5. Secure the accident scene
   ☐ barricade the area to prevent removal or defacement of possible evidence
   ☐ isolate potential witnesses

6. Collect and preserve the evidence
   ☐ make drawings of the area
   ☐ pick up, store, and label evidence
   ☐ if evidence cannot be removed from the scene or it is too large to bag, take pictures or make drawings, making sure to note the location on the drawing of the area
   ☐ make notes of any observations regarding the accident scene

7. Develop witness questions
   ☐ form open-ended questions based on initial observations and evidence collected

☐ include control questions to ensure the accuracy of the statistical data and to permit later evaluation of witness reliability

8. Interview the witnesses
    ☐ interview each witness separately
    ☐ find a suitable location
    ☐ be prepared to take notes on and/or record the interview
    ☐ take short notes as a memory device
    ☐ if a recording device is used, ask for permission to record before the interview begins
    ☐ watch for witness cues during the interview
    ☐ take into account personality types
        ● introvert
        ● extrovert
        ● suspicious
        ● prejudiced
        ● other personality traits
    ☐ notice any nonverbal messages
        ● body language
        ● voice changes
    ☐ establish initial communication
    ☐ make sure the individual understands that the purpose of the interview is for accident prevention not assigning blame
    ☐ note whether they were
        ● the injured party
        ● an eye witness
        ● an ear witness
    ☐ take an initial statement
    ☐ have the witness describe the incident in her own terms

☐ avoid interrupting the witness during the statement

☐ expand the interview for detail

☐ ask the developed open-ended questions

☐ space the control questions throughout this portion of the interview

☐ close the interview

☐ ask the witness for his suggestions on how the accident could have been prevented

☐ thank the witness for her time

☐ evaluate the witness statements

☐ develop a witness analysis matrix based on the control questions

☐ place the control question numbers on the horizontal axis

☐ put the witnesses on the vertical axis

☐ place an **x** in those columns where the witness has accurately answered the control question

☐ check credibility of individual testimony based on control questions

☐ high witness credibility is based on the number of accurate responses to the control questions

caution: just because a witness gives inaccurate answers to the control questions does not totally invalidate his testimony

9. If necessary, conduct follow up witness interviews

☐ let the witness know there is a gap or deficiency in a critical area of the investigation

☐ allow the individual to reconsider or reexamine her observations

caution: be sure to remain nonjudgemental

10. Synthesize the information gathered from the witnesses interviewed and evidence collected to determine the accident cause

11. Use the data gathered from the accident to perform an accident trend analysis

12. Make changes to operating procedures, equipment, and/or training based on the accident trend analysis

CONCLUSION

Several accident causation models have been presented in this review. Some of the theories focus upon people variables, on the management aspects, and others on the physical characteristics of hazards. Remember that the primary purpose and benefit of understanding accident causation is to recognize how hazards in the workplace can result in losses. By recognizing hazards and understanding how losses will result, the safety and health professional is in a better position to be proactive. Eliminating hazards before they result in losses is the proactive responsibility of everyone in an organization. The safety and health professional must find an effective means of communicating this concept to all members of an enterprise.

Safety professionals recognize that it is not always possible to identify and eliminate all hazards. Accidents may still occur in spite of a proactive safety program. It is at that point that an effective accident investigation program is of vital importance for the collection of critical data. Review the following material to identify the methods and benefits of an effective accident investigation program.

Questions

1. What are some of the benefits associated with the understanding of accident causation theory? Explain your answer.

2. What are the advantages of considering accidents as management problems?

3. What are the two factors associated with risk? Explain how these two factors impact upon the selection of accident controls.

4. Is it important to conduct accident investigations? Why?

5. What are the six key questions that should be asked during accident investigations? Explain the importance of these six questions to the health and safety professional.

References

Bird Jr., F. E., and Germain, G. L. 1992. *Practical loss control leadership*. Loganville, GA: International Loss Control Institute, Inc.

Bird, F. E., and Loftus, R. G. 1976. *Loss control management*. Loganville, GA: Institute Press.

Brauer, R. L. 1990. *Safety and health for engineers*. New York: Van Nostrand Reinhold.

Ferry, T. S. 1984. *Safety program administration for engineers and managers*. Springfield, IL: Charles C Thomas.

Ferry, T. S. 1988. *Modern accident investigation and analysis* (2nd ed.). New York: John Wiley and Sons.

Goetsch, D.L. 1993. *Industrial safety and health: In the age of high technology*. New York: Maxwell Macmillan International.

Grose, V.L. 1992, August. System safety in rapid rail transit. *ASSE Journal, 22,* 18-26.

Hale, A. R., and Glendon, A. I. 1987. *Individual behavior in the control of danger.* New York: Elsevier Science Publishers B.V.

Heinrich, H.W., Petersen, D., and Roos, N. 1980. *Industrial accident prevention.* New York: McGraw-Hill Book Company.

Kuhlmann, R.L. 1977. *Professional accident investigation: Investigative methods and techniques.* Loganville, GA: Institute Press.

Manuele, F. A. 1993. *On the practice of safety.* New York: Van Nostrand Reinhold.

Marcum, C. E. 1978. *Modern safety management practice,* Morgantown, WV: Worldwide Safety Institute.

Rasmussen, J., Duncan, K., and Leplat, J. (Eds.). 1987. *New technology and human error.* New York: John Wiley & Son.

Roland, H. E., and Moriarty, B. 1990. *System safety engineering and management* (2nd ed.). New York: John Wiley & Sons, Inc.

Chapter 6

Introduction to Industrial Hygiene

CHAPTER OBJECTIVES

After completing this chapter, you will be able to

- Define industrial hygiene terminology.
- List the responsibilities of industrial hygienists according to the "Code of Ethics."
- List and explain the activities associated with the industrial hygiene tetrahedron.
- Explain the process involved in occupational health hazard exposure.
- Identify the categories of occupational health hazard controls.
- Describe the modes of entry into the body for common health hazards.
- Identify the different forms of occupational health hazard contaminants.
- Determine the health risk of employees, resulting from contaminant exposure.
- Differentiate between voluntary and mandatory health hazard exposure limits.
- List the names of the four types of environmental stressors.
- Explain the concepts associated with occupational toxicology.

CASE STUDY

Employees at a Midwestern utility company complained of headaches, nausea and general fatigue. The building manager asked the corporate Health, Environment, and Safety Services support group for assistance. A team of three industrial hygienists was sent to the thirty-five year old facility to determine if there was an occupational health problem. Upon their arrival, and following discussions with building and department managers, they conducted a walk-through inspection. Employees, including those submitting complaints, were interviewed to determine the extent of the potential problem(s). Based upon the results of these interviews and discussions, an occupational health hazard evaluation plan was developed.

The evaluation plan called for extensive air and workplace environment contaminant monitoring. Temperature, humidity, dust, organic solvents, oxygen, carbon monoxide, carbon dioxide, radon, and a variety of other conditions or contaminants were measured. The team used the latest industrial hygiene monitoring instruments available to obtain and analyze the samples. These measurements were then compared to standards established by OSHA, NIOSH, EPA, ASHRAE (American Society of Heating, Refrigerating and Air Conditioning Engineers) and the company. This was performed to evaluate the concentration of contaminants found, as well as to determine if required or recommended human exposure standards had been exceeded.

Results of the monitoring and evaluation activities indicated that the facility had elevated levels of carbon monoxide, carbon dioxide, and organic solvents. Two conditions appeared to be contributing to the problems observed. First, the building had been constructed during the 1970s when owners were concerned with energy conservation. As a closed environment, the only replacement for the stale air was by the heating and air conditioning system. Windows could not be opened to supply air. To cut the costs associated with the heating and cooling of the facility, the building manager had reduced the amount of replacement air from outside the building. By increasing the amount of fresh air taken in from outside the building, the carbon monoxide and carbon dioxide problems were eliminated.

The second problem concerned the organic solvent concentrations that was found to be associated with the copy machine fluid. The copy machine, along with copy machine supplies, was kept in a converted closet. Employees working near the copy room were the individuals experiencing the symptoms associated with solvent exposure. The team recommended that the copy room have a separate ventilation system installed to remove solvent vapors before they could reach the affected workstations. It was also recommended that copy machine fluids be stored in a location where employees would not be exposed to the vapors.

Three months after the team submitted these recommendations, a follow-up inspection was conducted. Air samples indicated that concentrations of air contaminants were well within acceptable limits of all referenced standards. In addition, employee interviews indicated that all of the health symptoms previously reported had disappeared.

INTRODUCTION

When the OSHA law was enacted on December 29, 1970, the primary emphasis of OSHA was injury prevention in the occupational environment. During the 1980s and 1990s, OSHA has redirected its attention toward worker health. Employee health promotion and occupational illness prevention are the two primary concerns of a discipline called Industrial Hygiene.

Industrial Hygiene is defined as "the art and science of anticipation, recognition, evaluation, and control of physical, chemical, biological, and ergonomic hazards or stressors arising in or from the workplace, which may cause harm or induce discomfort or adverse health effects to workers or members of the community." Industrial hygiene is concerned with health hazards in the workplace. When a worker does not have enough oxygen to breathe, suffocation may be the health hazard of primary concern. During electroplating, degreasing, or painting operations, individuals may be exposed to toxic materials that can result in damage to the respiratory or circulatory system. In coal mines, quarries, or sand blasting rooms, workers can be exposed to dusts that can create numerous respiratory health problems.

Health hazards can exist in every occupational setting. It is the industrial hygienist who is given the task of examining the workplace, identifying potential health hazards, evaluating the magnitude of the health problem, and implementing control methods to eliminate the hazards if health hazards exist. According to the American Industrial Hygiene Association, an industrial hygienist is an individual who has been trained in engineering and/or the natural sciences and, through advanced studies, has acquired the knowledge and skills necessary to identify, assess, and correct health hazards in the occupational environment.

Because human life is so vulnerable when it is exposed to occupational health hazards, the American Academy of Industrial Hygiene has established a strict Code of Ethics for the Professional Practice of Industrial Hygiene. Individuals who meet the educational and experiential requirements listed above and successfully complete the examination requirements necessary to receive the designation Certified Industrial Hygienist (CIH), become members of the Academy and are expected to adhere to these standards. The purpose of this code and the responsibilities of industrial hygienists are presented below:

Figure 6-1. AAIH Code of Ethics

The American Academy of Industrial Hygiene
CODE OF ETHICS
For the Professional Practice of Industrial Hygiene

Purpose

This code provides standards of ethical conduct to be followed by industrial hygienists as they strive for the goals of protecting employees' health, improving the work environment, and advancing the quality of the profession. Industrial hygienists have the responsibility to practice their profession in an objective manner following recognized principles of industrial hygiene, realizing that the lives, health, and welfare of individuals may be dependent upon their professional judgment.

Professional Responsibility

1. Maintain the highest level of integrity and professional competence.

2. Be objective in the application of recognized scientific methods and in the interpretation of findings.

3. Promote industrial hygiene as a professional discipline.

4. Disseminate scientific knowledge for the benefit of employees, society, and the profession.

---

5. Protect confidential information.

6. Avoid circumstances where compromise of professional judgment or conflict of interest may arise.

Responsibility to Employees

1. Recognize that the primary responsibility of the industrial hygienist is to protect the health of employees.

2. Maintain an objective attitude toward the recognition, evaluation, and control of health hazards regardless of external influences, realizing that the health and welfare of workers and others may depend upon the industrial hygienist's professional judgment.

3. Counsel employees regarding health hazards and the necessary precautions to avoid adverse health effects.

Responsibility to Employers and Clients

1. Act responsibly in the application of industrial hygiene principles toward the attainment of healthful working environments.

2. Respect confidences, advise honestly, and report findings and recommendations accurately.

3. Manage and administer professional services to ensure maintenance of accurate records to provide documentation and accountability in support of findings and conclusions.

4. Hold responsibilities to the employer or client subservient to the ultimate responsibility to protect the health of employees.

Responsibility to the Public

1. Report factually on industrial hygiene matters of public concern.

2. State professional opinions founded on adequate knowledge and clearly identified as such.

INDUSTRIAL HYGIENE TETRAHEDRON

Industrial hygienists are responsible for anticipating, recognizing, evaluating, and controlling health hazards in the occupational environment. This list of industrial hygiene responsibilities is often referred to as the *Industrial Hygiene Tetrahedron*. The Industrial Hygiene Tetrahedron provides the strategy that directs the health protection program. By virtue of industrial hygienists' knowledge, skills, and experience, they must *anticipate* health hazards long before arriving at the employees' work environment. During this first phase, industrial hygienists may review facility schematics, purchase orders, material safety data sheets (MSDSs), standard operating procedures, or any other information that might be of assistance in focusing upon potential health hazards.

During the second phase of the Industrial Hygiene Tetrahedron, industrial hygienists perform qualitative and subjective observations in an attempt to *recognize* potential health hazards. Performing inspections to identify specific processes and methods in use at an operation, industrial hygienists identify potential worker exposure to environmental stressors. Environmental stressors include chemical, physical, biological, or ergonomic health hazards that pose the potential for injury, illness, or discomfort. Through the use of the human senses, industrial hygienists maintain an awareness of their reactions while involved in looking, listening, smelling, tasting, or touching the key components of the workplace. Based upon the anticipation and recognition phases, a preliminary strategy is established for performing the evaluation functions. See Figure 6-2 for a model of health risk assessment.

HEALTH HAZARD EXPOSURE

POINT OF ORIGIN→   PROPAGATION→   EMPLOYEE

Figure 6-2. A model of health risk assessment.

The third phase of the Industrial Hygiene Tetrahedron is the *evaluation* phase. Evaluation requires that hygienists use specific and sensitive field monitoring instrumentation to collect as much quantitative data as is possible and practical (*See* Figures 6-3, 6-4, and 6-5.) Samples are collected using approved and/or recommended monitoring methods. Depending upon the method used and the degree of accuracy desired, the samples may be evaluated at the site or sent to an approved laboratory for analysis. Based on the results obtained during the sampling activity, exposure calculations are performed to assess the concentration or amount of the contaminant present in the workplace. Considering the results obtained, the hygienist must then determine the extent of the health hazard exposure and compare these results with federal, state, or municipal regulations.

Once an accurate assessment of contaminant concentrations and health hazard exposure have been established, industrial hygienists must perform the most important responsibility of the Industrial Hygiene Tetrahedron. *Control* methods must be determined to eliminate, minimize, and reduce the health hazard. Industrial hygienists have three categories of health hazard control options available which, in the order of preference, include: (1) engineering controls, (2) administrative controls, and (3) personal protective equipment. The control method of choice is using engineering controls. This is the preferred protection method used to eliminate the hazard at the origin of the contaminant. By applying the principles of enclosure, encapsulation, or removal at the point of origin, industrial hygienists ensure that employees will have a reduced likelihood of exposure.

TYPES OF CONTROL

Controls can be defined as processes, procedures, or method changes that correct existing health problems and prevent or minimize the risk of health hazards in the workplace. Engineering controls are the method of choice because of their ability to totally isolate or eliminate health hazards. By eliminating health hazards at the point of origin, the occupational health and safety professional eliminates the propagation of the contaminate into the workplace environment, ultimately preventing employee exposure. Examples of engineering controls include the use of the contaminate into the workplace environment, ultimately preventing employee exposure. Examples of engineering controls include the use of ventilation systems to reduce the concentration of contaminants or enclosing and shielding hot work areas. Ventilation removes the hazard before propagation can occur. Enclosure places a barrier between the employee and the health hazard point of origin. In both examples, the employee is separated, and thus protected, from the health hazard in the workplace environment.

When it is impossible to implement engineering controls or when fabrication and construction of engineering controls will take time, administrative controls and personal protective equipment options may be required. Administrative controls are those health hazard control methods that are employee or process-management oriented. In other words, they are the control methods that the management of a facility has influence over through manufacturing method or employee work assignment activities.

Job rotation, moving employees from one workstation or task to another at regularly assigned intervals, is one example of an administrative control. For example, if noise is the workplace health hazard of concern, it may be possible to rotate the employee to a less noisy area during portions of the shift. This would reduce the overall concentration of exposure to noise for the entire shift. In this way, the overall exposure to the noise hazard is reduced, thus reducing the employees risk of hearing loss.

Other administrative control options include substituting less toxic materials used in the manufacturing and work processing, and establishing training programs that make the employee aware of the existing health hazard. The use of citric-acid-based solvents in place of a cancer-causing agent like carbon tetrachloride or the use of enamel-based paint in place of lead-based paint would be examples of the administrative control method referred to as substitution.

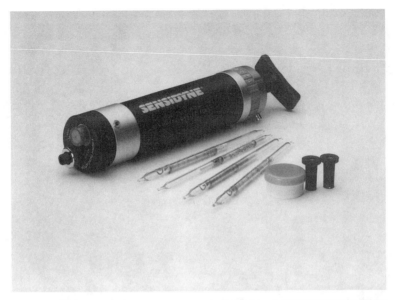

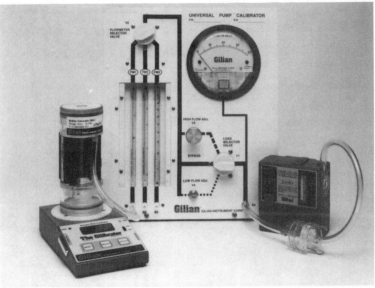

Figure 6-3. (a) An example of a piston air sampling pump with solid sorbent detector tubes used to obtain air samples for instantaneous detection. (b) Continuous air sampling pump and calibration equipment. (Photograph courtesy of Sensidyne, Inc., Clearwater, FL)

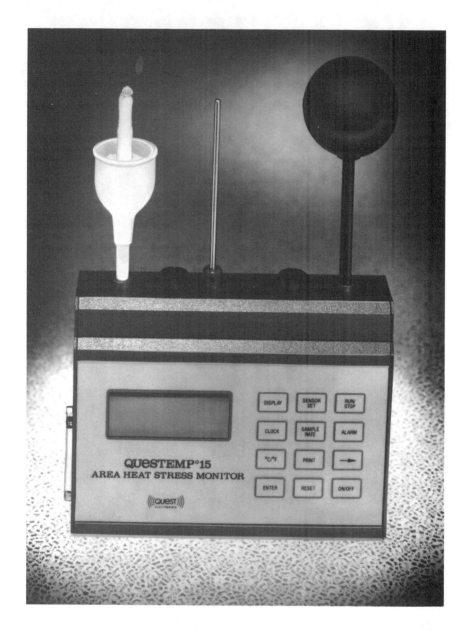

Figure 6-4. An example of an area heat stress monitor. (Photograph provided by and reprinted with the permission of Quest Technologies, Oconomowoc, Wisconsin)

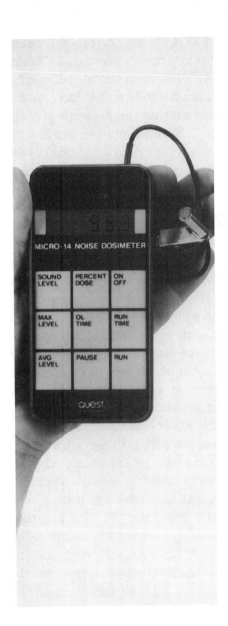

Figure 6-5. (a) Sound level meter with attached octave band analyzer. (b) An example of a personal noise dosimeter. (Photographs provided by and reprinted with the permission of Quest Technologies, Oconomowoc, Wisconsin.)

Personal protective equipment (PPE) is the third category of health hazard control available to the industrial hygienist or occupational health and safety professional. It is considered the last line of defense because the barrier separating the employee from the health hazard is worn by the individual. If this barrier fails, for what ever reason, the employee will come in contact with the contaminant. It is an unfortunate fact that for many occupational safety and health programs, PPE is the symbol of the safety program. However, the problem with PPE is that this control measure is only effective when used and maintained properly. The wrong piece of protective equipment can produce deadly results. (*See* Figures 6-6 and 6-7 for examples of Personal Protective Equipment.)

The main reason why administrative controls and PPE are not considered the preferred method for protecting worker health is that the health hazards still exist in the workplace and can present a risk. In many instances, both administrative controls and PPE can be circumvented. Employees, for example, may use the wrong respirator for a particular hazard. This mistake will result in the employee inhaling toxic vapors or dusts, increasing the risk of occupational disease. During a work site inspection on the midnight shift, employees were observed performing a lead paint spray operation while wearing a disposable gauze mask intended for low-level concentrations of dusts. In another inspection, employees with beards were using respirators. Both of these situations were dangerous to the health of those workers observed. In the first example, the employees had been trained in the importance of using respiratory protection but had mistakenly used the wrong equipment, assuming that it would protect them from all health hazards. In the second situation, the facility did not correctly implement the PPE program. Research has clearly demonstrated that leakage will occur when employees do not obtain good respirator face seals, and beards are just one cause of poor respirator face seals. Administrative controls and the use of PPE are important components of a health hazard protection program. They should not be considered the first choice of employee protection because of the inherent mistakes or misuse of these methods.

Figure 6-6. (a) An example of a full facepiece air purifying respirator. (b) An example of half facepiece respirator used with safety glasses and hard hat. (Photographs provided by and reprinted with permission of North Safety Products, Cranston, Rhode Island.)

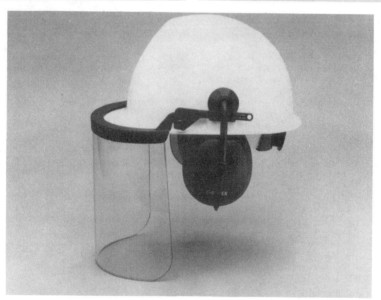

Figure 6-7. (a) Examples of laser eye protection. (b) Example of hearing protection. (c)Example of hard hat with face shield with Capmount Hearing Protection. (Photographs provided by and reprinted with the permission of Elvex Corporation, Bether, Connecticut.)

MODES OF ENTRY

The Industrial Hygiene Tetrahedron provides the health and safety professional with the health hazard recognition and control responsibilities that are required to implement and maintain an effective occupational health program. Knowledge of the way that contaminants can enter the employee's body helps in recognizing the potential health hazard risk, as well as determining the best control options that would prevent this exposure. Employee contacts with health hazards are often referred to as *modes* or *routes of entry*.

There are four common modes of entry: (1) inhalation, (2) ingestion, (3) absorption (external-through the skin or internal-through internal tissue), or (4) injection. Inhalation is probably the most common way that employees come in contact with health hazard contaminants in the workplace. As physical activity increases, the employee's body requires more oxygen in order to metabolize the fuels for use by the muscles performing the work. As the physical demands of the body increase, there is a corresponding increase in cardiopulmonary activity. Ultimately, as the individual inhales more deeply, additional contaminants are drawn into the body.

Ingestion is the mode of entry where the contaminant is taken into the body when people eat or drink. Ingestion exposures are often the result of poor housekeeping and/or substandard personal hygiene. For example, individuals can ingest a degreasing agent like toluene by not washing their hands before eating lunch. The sandwich consumed at the break can then be the means by which that solvent can enter the body. Housekeeping can pose an ingestion risk when food or beverages are placed on contaminated surfaces. Lunch rooms, where workplace dusts accumulate on eating surfaces, increase the risk of employees contaminating the food that they consume. Although this method of entry is slower then inhalation, concentrations can build up to lethal levels in the body.

Absorption is the mode of entry where the contaminant passes through the skin to produce negative health effects. Solvents or waste oils like toluene, xylene, polychlorinated biphenyls (PCBs), or stoddard solvent can damage the skin causing diseases such as dermatitis. Dermatitis has long been one of the leading causes of occupational illness.

The final mode of entry of concern to the health and safety professional is injection. For many, there is some confusion between absorption and injection. Contaminants entering the body by way of absorption diffuse through the skin tissue. Injection, on the other hand, is the introduction of the contaminant below the skin where diffusion has not occurred. Puncture injuries or the passage of contaminants through lesions of the skin are examples of injection.

For years injection was not considered a health hazard in most occupational environments. Injection routes of entry were associated with medical workplaces where accidental needle sticks were commonplace. With the increasing concern over AIDS and hepatitis B, injection routes of entry have been of increasing concern in first aid, emergency response, and medical treatment environments. Health and safety professionals should not be lulled into a state of complacency because they do not have medical personnel or facilities at their operation. Injections can occur whenever puncture wounds take place. For example, a contaminant may be present on a nail that an employee steps on resulting in the introduction of that contaminant into the circulatory system. Reports of the injection of a solvent under the skin during the cleaning of paint spray guns are another example of how injection-related entry of contaminants into the body may be more commonplace than initially thought.

No matter which mode of entry the contaminant takes, the health effects can be either local or systematic, depending upon the physical and chemical characteristics of the substance. *Local effects* are the tissue reactions of the body areas that come in direct contact with the contaminant. *Systemic effects*, on the other hand, are the target sites of the body that are not the initial contact locations, but because of their affinity for that particular substance, they readily absorb it. This reaction occurs often far away from the point of initial contact. If an employee were working with sulfuric acid and inhaled sufficient quantities, irritation of the upper respiratory system would be likely (irritation of the nose, bronchi, trachea, etc.). Inhalation of heavy metals, such as lead, would have systemic effects because of their tendency to not have a substantial impact upon the respiratory system. However, the primary danger is to the central nervous system where mental activities of the brain and psychomotor (eye-hand) coordination can be affected.

FORMS OF CONTAMINANTS

Contaminants can take a wide variety of forms. The following are some of the key terms associated with contaminant forms.

Aerosols: Liquid droplets or solid particulate dispersed in air.

Dusts: Solid particulates generated by crushing, grinding, chipping, or abrasion. Examples of dusts include: coal, wood, and sand.

Fumes: Solid particulates generated by condensation from a gaseous state. Fumes can include welding particulates.

Gases: A substance which is in the gaseous state at room temperature and pressure. Gases include oxygen, nitrogen, sulfur dioxide, and ammonia.

Mists: Suspended liquid droplets generated by the condensation from gas to liquid state.

Vapors: Gaseous phase of a material which is ordinarily a solid or liquid at room temperature. Vapors can include gasoline, solvents like toluene, xylene, and benzene.

Concentration for Classes of Contaminants

For substances including dusts, fumes, and mists, particulate toxicity is usually measured in terms of weight of the substance captured on the sampling media per unit volume of air crossing the surface of that media. This is typically represented in milligrams per cubic meter of air sampled. The volume of air sampled is determined by the flow rate of air volume passing through the sampling media over a fixed period of time. Fiber particulate toxicity, on the other hand, is measured as the ratio of fibers to the volume of air sampled, typically in cubic centimeters. Fibers such as asbestos, cotton, or fiberglass are collected on the surface of the sampling media as a unit volume of air crosses the media surface.

While particulates are primarily sampled to determine toxic effects upon workers, explosivity may also be of concern. Many of us are already aware of the grain dust explosions that can result when concentrations of particulates are sufficiently elevated and an ignition source is present. When monitoring to evaluate explosive hazards in the occupational environment, dust-to-air concentrations, as well as particle size and sources of ignition must be determined.

Gas and vapor toxicity is measured in units of concentration that compare the volume of the contaminant to the volume of air sampled. This volume-to-volume ratio is stated in parts per million or ppm. However, percent is used as the unit of measure when monitoring for flammable gases or vapors or oxygen deficient environments. This percent represents the ratio of the gas or vapor of concern versus the total volume of the air sampled.

During the evaluation phase, the occupational safety and health professional collects the sample on a sampling media. The safety professional will then measure the amount of the contaminant, such as volume of gas, weight of dust, or number of fibers. At the same time, the safety professional must determine the volume of air sampled. This is established by pre- and post-calibrating the sampling equipment used to establish the volume of air crossing the sample collection media during fixed units of time. Once again, the volume of air sampled is the product of the volume of air passing across the media during a given interval of time. The quantitative amount of the substance collected is then divided by the volume of air sampled and expressed in terms of ppm, percent, mg/M^3, or fibers per cc, depending upon the specific substance being monitored.

Determining the Health Risk to Employees Resulting from Contaminant Exposure

Once the concentration of the contaminant in the employees' work area has been established, the health and safety professional or industrial hygienist compares the results to established criteria. Established criteria has been set by two groups: OSHA and the American Conference of Governmental Industrial Hygienists (ACGIH). As previously discussed in earlier chapters, OSHA requirements are mandatory standards. This means that professionals must ensure that employee exposures fall below the specified limits or risk citations and penalties for failing to comply with these standards. The ACGIH standards are voluntary which means they are only recommended by panels of experts. Companies are not required to comply with the ACGIH standards, which tend to be more conservative (lower recommended concentrations). The OSHA health hazard standards of exposure are referred to as permissible exposure limits (PELs) while the recommended ACGIH standards are referred to as threshold limit values (TLVs). TLVs are published annually in the publication: *Threshold Limit Values and Biological Exposure Indices.* TLVs are defined in that publication as "the airborne concentrations of substances and represent conditions under which it is believed that nearly all workers may be repeatedly exposed day after day without adverse effect.

TLVs are based on the best available information from industrial experience, from experimental human and animal studies and when possible, all three. The basis on which the values are established may differ from substance to substance; protection against impairment of health may be a guiding factor for some, whereas reasonable freedom from irritation, narcosis, nuisance, or other forms of stress may form the basis for others" (ACGIH, 1994, p. 2).

There are three categories of TLVs:

Threshold limit value-time weighted average (TLV-TWA): The time-weighted average concentration for a normal 8-hour workday and a 40-hour work week, to which nearly all workers may be repeatedly exposed day after day, without adverse effects.

Threshold limit value-Short term exposure limit (TLV-STEL): The concentration to which workers can be exposed continuously for a short period of time without suffering: 1) irritation, 2) chronic or irreversible tissue damage, or 3) narcosis of sufficient degree to increase the likelihood of accidental injury, impaired self rescue, or materially reduced work efficiency and provided the TLV-TWA is not exceeded. The STEL is equal to 15 minute TWA exposure which should not be exceeded at any time during a workday even if the 8-hour TWA is within the TLV. There should be at least 60 minutes between successive exposures for a maximum of 4 exposures per day or per work shift.

Threshold limit value-ceiling (TLV-C): The concentration which should not be exceeded, even instantly.

When OSHA establishes contaminant exposure limits they are referred to as permissible exposure limits (PELs). PELs are typically expressed in terms of time-weighted averages or ceilings using the same definition and parameters as ACGIH's TLVs. OSHA permissible exposure limits are published in the tables found in Subpart Z of 29 CFR 1910. While health and safety professionals know that the OSHA PELs must be met to avoid citations, progressive companies attempt to meet the more conservative ACGIH standards. By meeting the TLV recommendations for maximum worker exposure to contaminants, health risks are reduced, and there is a greater margin of compliance in the event that health risk control methods are compromised.

ENVIRONMENTAL STRESSORS

Industrial hygienists categorize occupational health hazards into four categories referred to as environmental stressors: chemical, physical, biological, and ergonomic. Chemical stressors include substances such as solvents, acids, caustics, and alcohols. Physical stressors include ionizing radiation (alpha, beta, gamma, neutron, x-ray radiation), nonionizing radiation (infrared, ultraviolet, visible light, radio frequency, microwave, and laser radiation), noise, and temperature. Biological stressors include hazards such as bacteria, mold, fungus, and insect-related contaminants. Ergonomic stressors are the human psychological and physiological injuries or illnesses associated with repetitive and cumulative trauma, fatigue, and exertion.

A knowledge of work processes and an awareness of the environmental stressors allow safety and health professionals to anticipate and recognize potential health hazards. Questions that must be answered during preliminary investigations and inspections should include: (1) What environmental stressors are present in the facility? (2) Where are the points of origins of those environmental stressors? (3) What are the forms that those environmental stressors take? Are they dusts, vapors, gases, mists? (4) What are the work processes at the facility? (5) Have these work processes changed in any way? (6) What are the raw materials, intermediate, and finished products that are involved in the work process? Do they pose health risks? (7) What are the worker activities required to perform the required tasks? Do these tasks require frequent repetitive motions or excessive forces for long duration? (8) What control methods are currently being used?

Once these basic questions are answered, the safety and health professional must then evaluate the potential health hazards by determining the concentration and duration of exposure to the contaminant or stressor. The potential mode of entry or exposure of the stressor must be determined. Based upon the physical and chemical properties of the stressor, the probability of contact, absorption, inhalation, or ingestion must also be addressed. Then the rate of generation of the contaminant must be determined. The industrial hygienist must determine if, based upon the nature, toxicity, concentration, and duration of exposure, a significant health risk exists, thus creating the potential for causing injury or illness.

TOXICOLOGY

Toxicology is the study of the nature and actions of poisons and their effects upon living organisms. Toxicology plays an important role in occupational health and safety. Typically, substances are studied in animal experiments to determine their effects. The amount of the contaminant is presented to the organism in various concentrations and the animal's reaction is monitored. There are three important dose-response terms that are commonly referred to when discussing the toxicity of a substance:

Median Lethal Dose LD50: The dose at which 50 percent of a population of the same species will die within a specified period of time, under similar experimental conditions. This term is usually expressed as milligrams of toxicant per kilogram body weight (mg/kg).

Median Effective Dose ED50: The dose required to produce a specified effect in 50 percent of a population, for example a non-lethal end point.

Median Concentration LC50: The concentration of toxicant in air which causes 50 percent of a population to die within a specified time. This concentration is usually expressed as milligrams of toxicant in air (mg/m3) or on a volume basis such as parts per million (ppm). This is not a dose and is dependent on breathing patterns, deposition, clearance, retention, as well as numerous additional variables.

CLASSIFICATION OF TOXIC MATERIALS

There are numerous ways to classify toxic materials. In general, toxic materials are categorized according to their physical or chemical characteristics or the physiological organs that they target. The following groups of toxic materials are representative of the common methods used to classify toxic substances:

Chemical Classification: This classification uses the chemical structure, nature, and composition that a substance possesses. Examples of chemical classifications include aliphatic, aromatics, acids, alcohols, ketones, ester, and ethers.

Physical Classification: This method attempts to examine toxic agents according to the form in which they exist in the occupational environment. It includes gases, vapors, and mists.

The following definitions are from Appendix A of 29 CFR 1910.1200:

Asphyxiants: Substances that prevent the inhalation or use of oxygen resulting in deprivation and suffocation.
> Simple: Simple asphyxiants are those agents that displace oxygen in an atmosphere resulting in an "oxygen deficient" environment (defined by OSHA as an oxygen concentration below 19.5 percent). Methane, nitrogen, or hydrogen are examples of gases that can displace oxygen and reduce the concentration below the level necessary to support life.
> Chemical: Chemical asphyxiants are substances that prevent the uptake or transportation of oxygen because of their chemical reactions with target locations of the body. Asphyxiants such as carbon monoxide prevent the uptake and transportation of oxygen by the

hemoglobin while hydrogen sulfide paralyzes the area of the brain that controls respiratory activity thus reducing oxygen inhalation.

Carcinogens: A chemical is a carcinogen if it meets one of the three following requirements. It has been evaluated by the International Agency for Research on Cancer (IARC) and found to be a carcinogen or a potential carcinogen. The chemical is listed as a carcinogen or potential carcinogen in the *Annual Report on Carcinogens* published by the National Toxicology Program (NTP) (latest edition). The chemical is regulated by OSHA as a carcinogen.

Cutaneous Hazards: Chemicals which affect the dermal layer of the body such as the defatting of the skin, rashes, or irritations. Chemicals that cause these effects include ketones and chlorinated compounds.

Eye Hazards: Chemicals that affect the eye or visual capacity such as conjunctivitis or corneal damage. Examples of these chemicals include organic solvents and acids.

Hematopoetic Agents: Chemical agents which act on the blood to decrease the hemoglobin function or to deprive the body tissues of oxygen. Symptoms affecting the body include cyanosis and the loss of consciousness. Chemicals that cause effects such as these include carbon monoxide and cyanides.

Hepatotoxic Agents: Chemical agents that produce liver damage, such as liver enlargement and jaundice. Examples of these hepatotoxins include carbon tetrachloride and nitrosamines.

Irritants: A chemical, which is not corrosive, but which causes a reversible inflammatory effect on living tissue by chemical action at the site of contact.

Mutagens: A reproductive defect which is caused by chemicals resulting in chromosomal damage.

Nephrotoxic Agents: Chemical agents that produce kidney damage, such as edema or proteinuria. These chemical agents include halogenated hydrocarbons and uranium.

Neurotoxic Agents: Chemical agents that produce toxic effects on the nervous system, such as narcosis, behavioral changes, and decrease in motor functions. These chemical agents include mercury and carbon disulfide.

Reproductive Agents: Chemical agents which affect the reproductive capabilities, such as chromosomal damage (mutations) and effects on fetuses (teratogenesis). Examples of these chemicals include lead and DBCP (1, 2, Di Bromo-3-Chloropropane) which can lead to birth defects or sterility.

Respiratory Agents: Chemical agents that irritate or damage the pulmonary system. Symptoms of these agents include coughing, chest tightness, and shortness of breath. Chemicals within this class include silica and asbestos.

Teratogens: A reproductive defect which can be caused by certain chemicals that results in birth effects/defects on the fetus.

Hazard Communication Standard

With an effective date of May 23, 1988, the purpose of the OSHA Hazard Communication Standard (29 CFR 1910.1200) is to identify hazardous materials found in the workplace. Once identified, employers are required to inform employees of the substances that they work with, as well as methods to prevent exposure. This is primarily accomplished through container labeling, other warning forms, Material Safety Data Sheets (MSDSs), and training.

This standard applies to any substance that could be hazardous from industrial housekeeping and cleaning agents to process chemicals. If employees can be exposed to the substance(s) under normal work conditions or during emergencies, then they must be knowledgeable of the risks and necessary precautions to be taken to prevent health problems.

All substances in the workplace must be evaluated to determine if they are hazardous and pose a potential health or safety risk.

Manufacturers are required by this standard to label their products and provide information regarding physical and chemical properties of the substances. Hazardous materials that are in the workplace must be labeled identifying the specific hazards. The hazard warning must include words, pictures, signs, and health hazard information. Portable containers used during a shift do not have to be labeled if they are used by the individual transferring the product. However, it is always a good idea to label all containers to prevent accidental exposures. Labels that are destroyed or contain incorrect information must be replaced immediately.

Companies purchasing substances used during manufacturing activities obtain hazard information from the product's manufacturer in the form of a Material Safety Data Sheet (MSDS). MSDSs must be updated within three months of any change in health or protection information change. The employer must establish training and information programs for employees exposed to hazardous substances during their initial assignment to the job or when a new hazardous substance has been introduced. During this training, employees must be informed of the standard, the hazards present in their work areas and associated with their jobs, what MSDSs are and where they are kept, and details of the written program. In addition, methods for recognizing and detecting the hazardous substances, as well as ways to safely handle the product and protect themselves from hazardous substance exposure, must be covered to comply with this standard.

To ensure continuous compliance with the Hazard Communication Standard, department supervisors should be required to: (1) Conduct an annual survey of chemical inventories in their departments to determine the chemical name and product name, type and quantity of the substance on hand, and the types of containers in which they are stored. (2) Check the labels on all containers ensuring that they are labeled and in English. (3) Ensure that updated MSDSs are available for all hazardous substances in their department and stored in the designated area. (4) Assure that a written program is in place.(5) Conduct a periodic Hazard Communication refresher training to ensure employee knowledge, awareness, and emergency preparedness.

Material Safety Data Sheet
May be used to comply with
OSHA's Hazard Communication Standard,
29 CFR 1910.1200. Standard must be
consulted for specific requirements.

U.S. Department of Labor
Occupational Safety and Health Administration
(Non-Mandatory Form)
Form Approved
OMB No. 1218-0072

IDENTITY (As Used on Label and List)

Note: Blank spaces are not permitted. If any item is not applicable, or no information is available, the space must be marked to indicate that.

Section I

Manufacturer's Name	Emergency Telephone Number
Address (Number, Street, City, State, and ZIP Code)	Telephone Number for Information
	Date Prepared
	Signature of Preparer (optional)

Section II — Hazardous Ingredients/Identity Information

Hazardous Components (Specific Chemical Identity; Common Name(s))	OSHA PEL	ACGIH TLV	Other Limits Recommended	% (optional)

Section III — Physical/Chemical Characteristics

Boiling Point		Specific Gravity (H₂O = 1)	
Vapor Pressure (mm Hg.)		Melting Point	
Vapor Density (AIR = 1)		Evaporation Rate (Butyl Acetate = 1)	
Solubility in Water			
Appearance and Odor			

Section IV — Fire and Explosion Hazard Data

Flash Point (Method Used)		Flammable Limits	LEL	UEL
Extinguishing Media				
Special Fire Fighting Procedures				
Unusual Fire and Explosion Hazards				

Figure 6-8. Sample Material Safety Data Sheet (MSDS).

CONCLUSION

Occupational health hazards have assumed a more significant role for the safety professional's list of responsibilities. With increasing numbers of new chemical products being produced and new manufacturing methods being used, the industrial hygienist and occupational safety and health professional must remain vigilant. Employee health and well-being is dependent upon the industrial hygienist's ability to anticipate, recognize, evaluate, and control environmental stressors. It is a very complex and demanding area of occupational safety and health, but the rewards more than outweigh the demands.

Questions

1. What are some of the responsibilities of an Industrial Hygienist?

2. Why is knowledge of the industrial hygiene tetrahedron useful to individuals responsible for protecting the health and safety of employees?

3. What are three categories of health hazard controls? Provide two workplace examples for each category.

4. How is knowledge of contaminant modes of entry useful to the health and safety professional?

5. What is the difference between PELs and TLV? Explain why knowledge of these exposure measurements is important.

6. How are toxic materials categorized?

References

American Conference of Governmental Industrial Hygienists. 1994. *TLVs threshold limit values and biological exposure indices for 1994-1995.* Cincinnati, OH: American Conference of Governmental Industrial Hygienists.

American Conference of Governmental Industrial Hygienists. 1994. *Industrial ventilation: A manual of recommended practice.* 21st ed. Cincinnati, OH: American Conference of Governmental Industrial Hygienists.

Clayton, G.D. and Clayton, F.E., Eds.. 1991-1995. *Patty's industrial hygiene and toxicology, vols. 1-.* New York: John Wiley and Sons.

Confer, R. and Confer, T. 1994. *Occupational health and safety: Terms, definitions, and abbreviations.* Boca Raton, FL: Lewis Publishers.

Craft, B. F. 1983. Occupational and environmental health standards. In R. W. Rom, Ed., *Environmental and occupational medicine*. Boston, MA: Little, Brown & Company.

Loomis, T. A. 1978. *Essentials of toxicology*. Philadelphia, PA: Lea and Febiger.

National Institute of Occupational Safety and Health. 1973. *The industrial environment—Its evaluation and control*. Washington, D.C: U.S. Government Printing Office.

National Safety Council 1994. *Accident facts, 1994 edition*. Itasca, IL: National Safety Council.

Occupational Safety and Health Administration. 1993. *OSHA industrial hygiene technical manual*. Chicago, IL: Commerce Clearing House.

Occupational Safety and Health Administration. 1995. *Code of federal regulations, title 29 sections 1910.134, 1910.1000, 1910.1200*. Rockville, MD: Government Institutes.

Plog, B. A. (Ed). 1995. *Fundamentals of industrial hygiene* 4th ed.. Itasch, IL.: National Safety Council.

Chapter 7

Ergonomics and Safety Management

CHAPTER OBJECTIVES

After completing this chapter, you will be able to

- Define ergonomic terminology.
- List and describe the components of the Operator-Machine System.
- Explain the role of anthropometrics when solving ergonomic problems.
- Explain the role of biomechanics when solving ergonomic problems.
- List the categories of workstations.
- Identify when selected types of workstations should be considered the design of choice.

CASE STUDY

Employees at a small manufacturing facility were experiencing a high frequency of back injuries while manufacturing and finishing air compressor tanks. The management of the plant was at a loss as to what to do about the rising workers' compensation premiums. Faced with the loss of their compensation insurance coverage, they contacted a Certified Professional Ergonomist (CPE) to conduct an analysis of the workplace.

The consultant first reviewed various safety and health records including OSHA 200 logs, nurse's logs, turnover records, and Workers' Compensation claims. Health interviews were then conducted. Following these basic ergonomic assessment activities, workplace layouts and critical design measurements were recorded. The consultant also videotaped various work locations in the plant where the majority of ergonomic problems were reported. Using the video tapes, the consultant analyzed each job observed by breaking them down into major behavioral steps. Each step was further analyzed for

- body motion,
- activity frequency and duration,
- pace issues, and
- lifting-related risk factors.

Lifting tasks were studied using the NIOSH lifting guidelines. Factors such as the weight of the objects being lifted, the distance that the objects were held away from the employee's body, and where objects were located at the beginning and end of the lift were some of the measurements taken. Individual risk factors were also studied along with environmental factors such as temperature, lighting, noise, and the other environmental stressors that can influence human performance effectiveness.

All of these analyses pointed to two locations where manual material handling was contributing to the back injury epidemic. Most of the injuries occurred during storage activities in the warehouse, as well as in the paint shop where tanks were loaded onto an overhead hook conveyor system.

The consultant, working with management and the company safety committee, developed several strategies for reducing the frequency of back injuries. Placing the tank storage pallets on scissor lifts, reducing bending and lifting activities, and using portable jib cranes so manual lifting was not required were some of the suggestions applied in the paint shop. In the warehouse, placing finished product at floor level instead of on shelves and eliminating the full-arm extended reaching and lifting postures reduced those risk factors. Through implementation of these strategies, the company saw improved productivity, improved employee morale at the two locations, and a 63 percent reduction in the frequency of back injuries.

INTRODUCTION TO ERGONOMICS

Scarcely a day goes by where one doesn't read or hear about ergonomics. Manufacturers of furniture, appliances, automobiles, and even toothbrushes are extolling the virtues of their ergonomically designed products. Back injuries and carpal tunnel syndrome are topics constantly raised in both work- and non-work-related gatherings. Ten years ago, people may have never heard of the term "ergonomics," but it has become a buzzword of the 1990s.

Ergonomics has obtained this status because it is a significant health and safety issue. In terms of medical costs, employee days lost from work, and human suffering, ergonomics has become a serious issue for most companies. Ergonomics is the leading occupational illness in the workplace and safety professionals are spending more of their time trying to solve ergonomic problems. Successfully eliminating ergonomic problems in the occupational environment requires that safety professionals develop systematic strategies to correct this problem.

There are several specific ergonomic activities that safety professionals perform as part of their strategy for eliminating ergonomic hazards. First and foremost is an examination of mental and physical job demands. When potential ergonomic problems have been identified, health and safety practitioners perform task analyzes to determine if the job demands exceed human capabilities. Knowledge of human physiological dimensions (anthropometrics) and movement (biomechanics) is an important part of this approach. Safety professionals use movement categories, based upon anthropometric and biomechanical classifications, to monitor the actions of employees while performing their tasks. Trends are then calculated to determine which motions, forces, or other conditions could be contributing to ergonomic problems.

Operator-machine system analysis is another part of the ergonomic evaluation process. This analysis examines factors associated with the people, environment, equipment, machinery, and layout of the workplace. The operator-machine system approach directs safety and health professionals toward an understanding of the causes of ergonomic problems, as well as the development of potential solutions for eliminating ergonomic hazards. This is accomplished by studying the interaction of operator-machine factors during task performance activities.

The final and most critical phase of this process is the elimination of ergonomic hazards. Control measures are designed by safety professionals to do away with the conditions that cause the ergonomic problems. Control measures may include engineering controls (redesigning the workstation) or administrative controls (changing the methods that workers use to perform their jobs).

This chapter is intended to provide the reader with some of the tools useful for recognizing and evaluating ergonomic hazards in the occupational environment. The tools and techniques briefly discussed in this introduction will be presented to help the reader recognize ergonomic hazards. In the last section, hot ergonomic topics will be reviewed. This "hot topic" review focuses on an examination of workstations, repetitive motion, and lifting.

DEFINITION OF THE TERM "ERGONOMICS"

The term *ergonomics* is based on two Greek words: *ergos* meaning "work," and *nomos* meaning "the study of" or "the principles of." In other words, ergonomics refers to "The laws of work." Ergonomics is the discipline that examines the capabilities and limitations of people. There are several fields of study that contribute to the ergonomics discipline. Together, these numerous fields are providing and compiling information about human characteristics and their connection with workplace tools, materials, or facilities. The discipline of ergonomics then applies this information to the design of working and everyday living environments.

MULTI-DISCIPLINARY STUDY OF ERGONOMICS

To be successful, ergonomics is dependent on several disciplines which study and apply information concerning human capabilities and limitations. It is also dependent on several disciplines which study and apply information about environments where people live and work. As a consequence of this broad area of study which overlaps many other fields of study, the understanding of ergonomic principles and applications has resulted in a multi-disciplinary effort.

The ergonomic multi-disciplinary approach involves accumulation of knowledge from such fields of study as sociology, psychology, anthropology, anatomy, physiology, chemistry, physics, mechanics, statistics, industrial engineering, biomechanics, and anthropometrics. Ergonomic principles and practices are then applied to the industrial environment through activities associated with human factors engineering, industrial engineering, occupational layout and design, product design, safety engineering, occupational medicine, or industrial hygiene.

ERGONOMICS OBJECTIVE

The primary objective of ergonomics is the improvement of human health, safety, and performance through the application of sound people and workplace principles. The goal of safety and health professionals should be the application of ergonomics as a safety function that is sought out as a service to the organization. Members of production management should be "ideally" seeking out the safety function for improvement of work group productivity and efficiency.

Because the primary objective of ergonomics is to improve productivity and efficiency, ergonomic services should not be viewed as an add-on activity. Ergonomists or individuals assigned those duties can best serve the organization as a part of a team. The "ergonomics team" would systematically analyze workplace layout and design, analyze job requirements from a worker capability and limitation perspective, and recommend improvement of the production process. Under ideal situations, these activities would be performed proactively to eliminate problems prior to the start of work-related activities. To accomplish this goal, ergonomic activities must be integrated into both the safety and production process. It should not be considered a separate area of responsibility.

APPLYING ERGONOMICS

The industrial hygiene triad, an occupational health-oriented, problem-solving approach, is a useful method to ensure an effective process for applying ergonomic principles in the workplace. This requires that a recognition, evaluation, and control strategy be applied.

Recognition of ergonomic hazards usually involves the search for symptoms. Physiological stresses and muscular strains, psychological stresses, and general complaints or discomfort are often symptoms of ergonomic problems. By recognizing these symptoms when they appear, safety practitioners need to locate where potential ergonomic hazards exist. Through preliminary observation of the workplace and interviews with key personnel, safety practitioners determine where employees are exposed to potential ergonomic hazards. If they believe that the symptoms are a result of ergonomic hazards based on these preliminary measures, then it would be necessary for safety practitioners to initiate evaluation activities.

Ergonomic *evaluation* activity implies the collection of information to help determine if a problem exists. Evaluation can determine both the extent of the problem, as well as provide the direction required to identify options for the elimination of those problems. Review of written records including OSHA 200 logs, first aid or nurses logs, and workers' compensation records should be considered as the starting point in the evaluation process.

If safety practitioners notice significant numbers of ergonomic injuries and illnesses, more detailed evaluation activities must be performed. Back injuries, carpal tunnel syndrome, tendinitis, tenosynovitis, and muscle strains or sprains are just some of the injuries and illnesses that indicate a need for further analysis. Written records such as the OSHA 200 log contain the name of the individual involved in the incident, the individual's job title, and the job's location at the facility. This information can be gathered and trends observed. Follow-up activities are then started for those jobs or in those locations where the greatest frequency or most severe incidents have occurred. Detailed field surveys such as personnel interviews, supervisor interviews, workplace observations, and employee health surveys are some of the follow-up evaluation activities that provide practitioners with added insight into the extent of the ergonomic problems at their facility.

The final stage of the evaluation process requires safety practitioners to conduct job or task analysis activities. Frequently, tasks are videotaped to permit a detailed evaluation of key ergonomic risk factors. Videotaped jobs are broken down into discrete behavioral steps. Each step is monitored and frequency measures recorded.

Factors studied during each step of the job include:

- the frequency of potentially harmful motions
- the length of time that the same task has to be performed
- the pace that employees must maintain during the shift
- the internal muscle forces that employees exert to perform tasks
- the external forces exerted on employees (such as the weight of objects that are carried)

With this type of job-related data in hand, control measures can be determined. Engineering *controls* such as ergonomically correct workstation layouts and workplace designs can be developed. If engineering controls are not possible, administrative controls should be considered as the next option for safety practitioners. Job rotation, requiring employees to perform several different jobs during a shift, is a useful administrative control method for eliminating workers from having to perform the same repetitive tasks. Another administrative control option would be to require two individuals to lift specific products that exceed company manual-material-handling recommendations.

This industrial hygiene triad of *recognition, evaluation,* and *control* has often been the process that safety practitioners have used to solve ergonomic problems. Specific tools and techniques, however, must be applied within this process to effectively pinpoint the presence of ergonomic hazards. One of the techniques frequently used is the operator-machine (or equipment) system of ergonomic analysis. Once the ergonomic hazards have been identified, the operator-machine system is used to provide the practitioner with a methodology to specify the exact workplace cause or causes of the problem.

OPERATOR-MACHINE SYSTEMS

When potential ergonomic problems have been recognized and superficially evaluated for specific jobs or locations, occupational health and safety professionals apply the operator-machine system to gather greater details. This system categorizes the major conditions that contribute to ergonomic problems.

The operator-machine system is made up of three components:

- People (the operators of equipment, machinery, vehicles, or tools)
- Machines (the tools, machines, vehicles, equipment, and facilities)
- Environment

This system provides safety professionals with a systematic procedure to study the three categories of causes that can result in ergonomic hazards existing in the occupational setting.

People Variables

The people or operator variables associated with this system are composed of the human factors that contribute to the ergonomic problem. They include psychological capabilities and limitations; physiological dimensions, capabilities and limitations; and psychosocial factors. One or more of these conditions can have the potential to cause injuries or illnesses in the workplace.

The people component of the operator-machine system focuses upon the psychological factors that can affect worker performance. *Psychology* is the science that studies human behavior. From an ergonomic perspective, psychological factors contribute to a wide variety of potential ergonomic problems. A few of the psychological causes of ergonomic problems include: memory, attention, fatigue, boredom, job satisfaction, future ambiguity, and stress.

Extensive data examining psychological capabilities and limitations have found that employee mental health has a profound effect upon the safety, quality, and productivity of an organization. For example, the stress associated with co-workers witnessing an accident could affect their personal safety. Observing a friend getting seriously injured can be both stressful and distracting. This, in turn, could result in their involvement in a future accident. Another example is what often occurs when traffic accidents take place. Passers-by tend to stare at the accident scene. From an ergonomic perspective, people are less attentive when their attention is divided between several tasks or activities. As a consequence, there is an increased likelihood of more accidents occurring during these situations. For more information on psychological factors, refer to Chapter 11.

Besides the psychological factors, the people component of the operator-machine system requires an understanding of human physiology. *Physiology* is the branch of the biological sciences concerned with the function and process of the human body. From an ergonomic perspective, the understanding of human physiology is vital if safety practitioners are to recognize when workplace demands exceed human capabilities. Physiological capabilities and limitations include both the structure, strength, and movement of anatomical components. These are studied in the disciplines concerned with anthropometrics and biomechanics.

Physiological factors can include the study of the human body at the cellular level. Neurological activity associated with light stimulating the retina when a warning signal illuminates is such an example. In addition, it can include such complex phenomena as gross motor functioning when studying how various muscle groups of the body tense and relax to provide movement and balance while lifting and carrying an object.

Psychosocial factors, the third element of the people component of the operator-machine system, are of great importance to any safety practitioner responsible for occupational ergonomics. The term psychosocial is derived from the term social psychology. It refers to an individual's behavior in a group environment. From an ergonomic perspective, topics examined in social psychology include attitude formation, attitude change, leadership styles, power and influence, conflict, occupational stress, organizational structure, employee motivation and organizational reward systems.

All of these psychosocial factors can influence employee performance in the social context of an organization. For example, the future uncertainty of a corporate takeover can influence worker morale and performance. Questions associated with the likely loss of jobs raises employee concerns over job security, forced early retirement, relocation and moving, as well as possible new job assignments. These concerns become distractions which again reduces the attention to detail of an assigned task.

It must be pointed out that these psychological, physiological, and psychosocial elements can occur either in isolation or in combination with one another. An example is shift work. Shift work is the term that refers to those staffing work schedules that are performed during non-traditional employment hours of a day. Shift work has been defined as "any regularly taken employment outside the day working window, defined arbitrarily as the hours between 7:00 a.m. and 6:00 p.m." (Monk and Folkard, 1992). It can refer to working on a fixed shift where individuals maintain the same schedule like a "cat eye" shift (the "cat eye" shift is a term used in coal mining where miners typically work a fixed schedule from 11:30 p.m. to 7:00 a.m.). Shift work also refers to rotating shifts where employee schedules are changed on a regular basis. An example of a rotating shift schedule would be where employees work midnight to 8:00 a.m., for one week, 8:00 a.m. to 4:00 p.m., the next week, and 4:00 p.m. to midnight the third week. Any time individuals are required to work non-traditional hours, employers are faced with the ergonomic problems associated with shift work.

Shift work physiological hazards and strain are those created when the body attempts to fight its natural circadian rhythm. The circadian rhythm is a biological process which is used to explain that humans are designed to be active and awake during daylight hours and to rest or sleep during hours of darkness. Susceptibility to fatigue increases when people have work schedules that conflict with this rhythm. From a medical perspective, physiological strain is manifested by an increased number of cardiovascular and gastrointestinal problems for individuals required to perform shift work (Rutenfranz et. al., 1977).

Psychological factors associated with shift work include the stress and depression associated with working unusual schedules. Individuals performing shift work have been found to be less effective in terms of their cognitive abilities and attentiveness on the job. In addition, safety records for shift workers have been found to be worse than day-shift employees (DeVries-Griever et. al., 1987).

The psychosocial factors associated with this problem are primarily interpersonal in nature. Often, workers will not be able to interact with spouses and friends for long periods of time, creating a void in their social support system. They cannot share the sorrows or joys of their accomplishments because they have lost contact with the important individuals in their lives. This loss of emotional support can result in marital difficulties and, ultimately, elevated frequencies of divorce.

Machine Variables

Machine or equipment variables examine workplace layout and design. It includes such factors as the position of the displays and controls associated with the operation and monitoring of machinery. Manufacturing workstation variables can include the distance of part storage bins from the worker, the distance of the conveyor systems from the worker area, or the table height of the work surface itself. Workstation layout can also be an area of concern in offices. Furniture design or layout of computer components are machine/workplace factors that can create ergonomic hazards. Conditions such as the back support provided by a chair, the height of the keyboard surface, and the position of the monitor would be evaluated by safety professionals to ensure that ergonomic hazards were not present. This equipment, machinery, or workplace variable also includes tools and equipment used, layout of aisles and walkways and a variety of other factors. Hand- and power-tool characteristics such as handles, shape, vibration, and methods of use have the potential for contributing to the ergonomics problem. Refer to the discussion concerned with classification of workstations found later in this chapter for more information on evaluating the machine elements of the operator-machine system.

Environmental Variables

Environmental variables, those factors examined in the chapter on Industrial Hygiene, include temperature, lighting, noise, vibration, humidity, and air contamination. All of these conditions can result in elevated ergonomic risks. For example, conditions such as hot temperatures and high levels of humidity create the potential for a wide variety of physiological and psychological problems. Fatigue often accompanies these conditions, which ultimately reduces the worker's mental capability to focus on a task. Health concerns such as heat stress, heat exhaustion, and heat stroke are additional dangers associated with extreme temperature and humidity conditions.

Light and workplace illumination is another risk factor that has the potential to cause ergonomic problems in a wide variety of work environments. Computer glare on a monitor in an office environment can cause eye fatigue and strain problems. Insufficient illumination in a warehouse can resulting in slipping and tripping incidents. Excessive illumination or glare can result in operator errors while using heavy equipment outdoors. Environmental variables should always be evaluated when analyzing the ergonomic hazards in the workplace.

Upon review of the elements of the operator-machine system, there are numerous factors that safety professionals must examine to identify and evaluate ergonomic causal factors. In some instances, even more in-depth information is required to pinpoint the causes of specific types of ergonomic problems. When physiological factors of the operator-machine system element are of concern, an in-depth evaluation would require safety practitioners to analyze tasks from an anthropometric and biomechanical perspective.

ANTHROPOMETRICS

Anthropometrics is the measurement and collection of the physical dimensions of the human body. These body measurements are used to improve the human fit in the workplace. These measurements can also be used to determine problems that exist between old facilities or equipment and the employees using them.

There are two types of anthropometrical dimensions that are useful for the study of human physiology and its affect on workplace layout and design: (1) *structural* or *static* anthropometry, and (2) *functional* or *dynamic* anthropometry. The body measurements and dimensions of subjects in fixed standardized positions are referred to as structural or static anthropometrics. Some common structural anthropometric measurements include: stature (height), sitting height, body depth, body breadth, eye height sitting or standing, knuckle height, elbow height, elbow to fist length, and arm reach.

Functional or dynamic anthropometry is the body measurements and dimensions taken during the performance of various physical activities. These movements may be required for the performance of particular types of tasks. Some of the frequently used functional measurements include crawling height, crawling length, kneeling height, overhead reach, bent torso height, and range of movement for upper-body extremities.

Anthropometric measurements help designers determine furniture or workplace layout requirements based on typical human body sizes. Prior to the inception of ergonomics and anthropometrics, many machines and workplaces were designed for the average employee. Unfortunately, average statistical measurements represent less than one percent of the normal distribution of body measures. For example, if a standing workplace were designed for an average American male (5'8"), 99 percent of the American population would not be properly fitted for this workplace.

Consider the seating and controls engineered for automobiles. Years ago, seats in cars could be adjusted forward and backward to accommodate the leg length and pedal reach requirements of the driver. Height adjustments, however, were not typically available. Cars were designed for the average person in our population. This condition caused short individuals to look through the steering wheel while tall individuals would strike their heads against the roof.

Today, ergonomists use anthropometric measurements that include at least 90 percent of the population. Most workplace designs attempt to achieve this goal by including people dimensions between the 5th and 95th percentiles. Normally, this is accomplished by providing workplaces that are adjustable.

From this discussion, it should be noted that percentile statistics are used extensively in anthropometrics. They are used to represent the number of individuals with measurements less than or equal to the dimensions of interest. If someone had an anthropometric measurement that placed him in the 95th percentile, that would mean that his measurement was as large or larger than 95 percent of the population. Consider the height (stature) of an American population of 50 percent males and 50 percent females. The 5th percentile of this population would be made up of those of a height of 60.8 inches while the 95th percentile would include individuals of a height of 72 inches. If someone were six feet (72 inches) tall, she would be in the 95th percentile. This would indicate that she were as tall or taller than 95 percent of the entire population. Ninety percent of this male/female population could then be accommodated for height if the workplace were designed to be adjustable for people between these two measurements of 60.8 and 72 inches.

What does this actually mean when looking at a workplace? Suppose that a safety professional is interested in eliminating ergonomic problems that result from workstations that are not correctly designed for their employees. Taking anthropometric measures and readjusting the workstation could eliminate the ergonomic design problems.

One example of this approach might be back injuries associated with lifting boxes onto a shelf. According to the recommendations associated with the NIOSH lifting equation, *knuckle height standing* at the start of a lift produces the least amount of stress upon a worker's back. In addition, if you can minimize the destination (height of the shelf) where the box has to be placed, this would also reduce back stress. Thus, back muscle stress could be greatly reduced if the boxes could be removed from a scissor pallet at knuckle height and stored at knuckle height.

In this example, safety professionals could take employee measurements or refer to anthropometric data for knuckle height standing. If professionals refer to the anthropometric charts, they would observe that male and female employees would have a knuckle height range of 25.9 inches (at the 5th percentile) to 31.9 inches (at the 95th percentile). What is the solution to this ergonomic problem? In one word, *"adjustability."* By having an adjustable scissor pallet and by providing a platform for shorter employees, this height requirement could be met. This would mean that 90 percent of the workforce would not have to bend to lift and store these products.

BIOMECHANICS

Those who have taken a physics course have learned that mechanics is the science of motion and force. Biomechanics is the human equivalent of this concept. *Biomechanics* can be defined as the study of the mechanical operation of the human body. It is the science of motion and force in living organisms. In biomechanics, the function of the body components is monitored and job requirements modified to lower internal and external stresses. It is the musculoskeletal system of the body that provides the foundation data for the study of biomechanics. The measurement and mechanism of body movements are important to safety professionals.

Like anthropometrics, there are two types of biomechanical measurements:

- **Dynamics**- the study of moving bodies.

- **Statics**- the study of bodies remaining at rest (equilibrium) as a result of forces acting upon them, and

These two measures provide the safety professional with the ability to determine how moving body components contribute to ergonomic injuries. For example, factors such as extension and force applied by the arm and leg muscles while pushing a hand truck might be a biomechanical area of concern. In addition to these areas of interest, safety professionals would also study the load on the hand truck, the friction coefficient of the walking surface, and other external forces that could affect employee performance.

In biomechanics, the measurement of primary concern is force. This is especially true as force relates to loads and stresses on the body. *Force is defined as that which can cause an acceleration of matter.* In biomechanics there are two categories of force that create motion of biological matter or, in everyday terms, movements like walking or lifting. These two categories are (1) *load*--the external forces upon a structure or organism, and (2) *stresses*--the internal forces generated in the structure as a result of loading.

In the human body, all movement is made possible by the application of load and stress to biological levers. With an awareness of joints, bones, and muscles, biomechanics provides safety professionals with an understanding of how the musculoskeletal levers of the body are designed to work. Knowing how the body is designed to naturally move, professionals can identify and eliminate the unnatural movements that can result in ergonomic problems. Evaluation activities such as monitoring frequency and duration of movement or examination of postures and positions can be initiated to determine the level of ergonomic risk. Internal and external forces should also be a part of this evaluation. These activities are performed as a part of a job or task analysis.

When twisting or other unnatural movements of these biological levers are observed during a task analysis, they should serve as warning signals to the safety professional. Ease of work activity or biomechanical advantage is only possible when the weight is held and moved using the best posture and body position. These best postures and movements are our natural physiological movements. Unnatural movements and postures will eventually result in ergonomic injuries

CLASSIFICATION OF BODY MOVEMENT, POSTURES, AND POSITIONS

Safety professionals can effectively determine the level of ergonomic risk by performing ergonomic task analyses. As mentioned previously, a job or task analysis requires the breakdown of an activity into discrete behavioral steps. Once the steps have been sequentially developed, a movement, force, or other systematic analysis can be conducted. If a movement analysis is conducted, safety professionals must then logically break down each step into the movements observed. Movements for each step are then recorded and used for calculating repetition rates and duration. This is an important activity because ergonomic problems have been correlated with frequency and duration of biomechanical motions.

There are a number of systems that can be used to classify and evaluate body movements. The two movement classification methods are physiological and operational categories. Physiological classification systems examine the way body parts move. Operational classification systems group body movements by the particular work activity being performed. Categories of physiological and operational movements will be closely examined.

Physiological Categories of Movement

Listings of body movements are used to classify the biomechanical movement of the individuals performing a task. The following is a cursory list of physiological movements with a brief description of each term (Kroemer, Kroemer, and Kroemer-Elbert, 1990). Refer to Figures 7-1, 7-2, 7-3, and 7-4 for clarification of the movements represented.

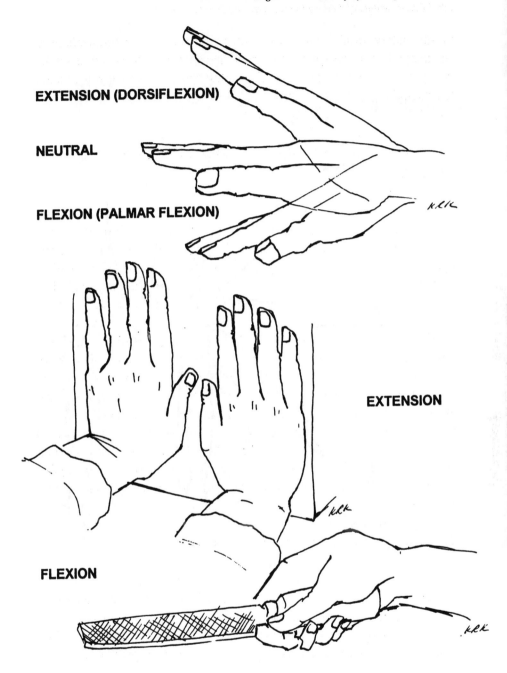

Figure 7-1. Examples of wrist flexion and extension.

ABDUCTION

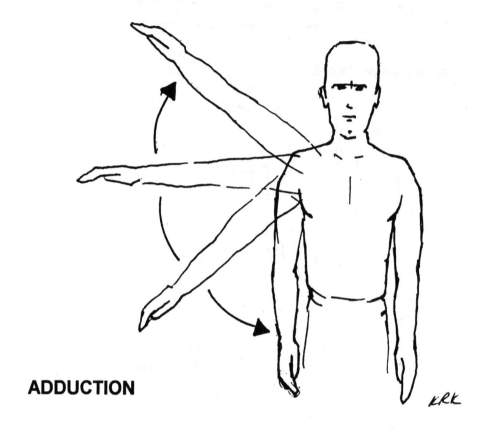

ADDUCTION

Figure 7-2. An example of arm abduction and adduction.

PRONATION   NEUTRAL PLANE   SUPINATION

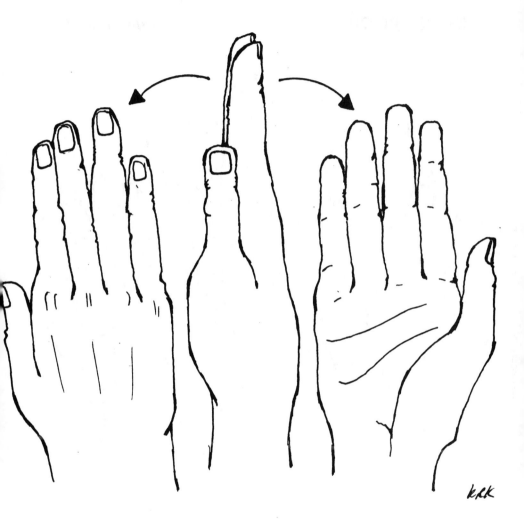

Figure 7-3. An example of wrist supination and pronation.

NEUTRAL PLANE

ULNAR DEVIATION   **RADIAL DEVIATION**

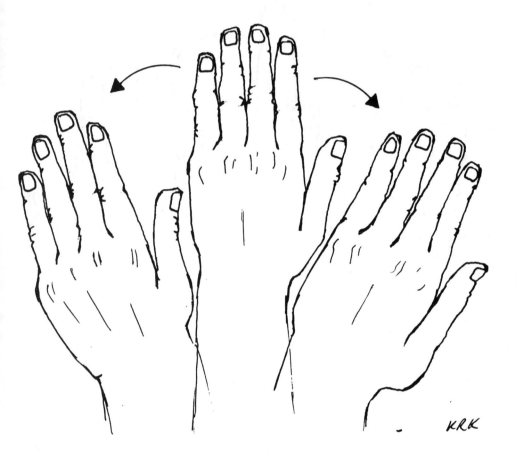

Figure 7-4. An example of hand neutral plane and deviation positions.

Abduction/Adduction

Abduction: The movement of a body part away from the center plane of the body. Lifting the arm outward away from the body is an example of abduction.

Adduction: The opposite of abduction. The movement of the body part toward the center plane of the body. Lowering the arm toward the body is an example of adduction.

Circumduction

Circumduction: Rotary movements which circumscribe an arc. Swinging the arm in a circle is an example of circumduction.

Flexion/Extension

Flexion: The movement of a joint that decreases the angle between the bones. Bending the arm at the elbow such that the hand moves closer to the upper arm region is one example.

Extension: The opposite of flexion. The movement of a joint that increases the angle between the bones. Straightening the arm is an example of extension.

Neutral Plane/Deviation

Neutral Plane: The normal and low-stress position of segmental physiological components. Maintaining the hand, wrist, and forearm at a 180-degree angle or in a straight and linear plane or dropping the hand, wrist, forearm, elbow, and upper arm at one's side are examples of neutral planes.

Deviation: The movement or position of a body part away from the neutral plane. Bending the wrist with the hand bent toward the thumb is referred to as *radial deviation*. Bending the wrist with the hand bent toward the small finger is referred to as *ulnar deviation*.

Rotation

<u>Rotation</u>: A movement in which a body part turns on its longitudinal axis. The turning of the head or arm is an example of rotation.

Supination/Pronation

<u>Supination</u>: The turning of the forearm or wrist such that the hand rotates and the palm is facing upward.

<u>Pronation</u>: The opposite of supination. The turning of the forearm or wrist such that the hand rotates and the palm is facing downward.

Operational Categories of Movement

Operational Classification of movement refers to the task being performed by the operator at the time of the job observation. The following is a list of the terms used to represent the operational classification of movements.

Positioning: This classification involves moving an object and corresponding extremity from one position to another. An example of positioning would be reaching for a bolt stored in a bin to the right of an employee.

Continuous movement: A single movement involving muscle control to adjust or guide a machine or other piece of equipment. An example of continuous movement would be the steering of a forklift.

Manipulative movement: The handling or assembling of parts. This movement classification is usually limited to hand or finger movement. An example of manipulative movement would be the assembly of component parts.

Repetitive movements: These are the same movements which recur over and over. Hammering or using a screwdriver would be examples of repetitive movements.

Sequential movements: A series of separate movements that are joined together in a specific order to complete a given task. Reaching for a tool with the right hand, grasping a component in the left hand, moving the two hands toward one another, and adjusting the component using the tool are examples of sequential movements.

Static Movements: Maintaining the position of a body member in order to hold something in place. Though movement may not be involved, the muscles are required to maintain the steady position of the object. Holding a board or plaster board in place on the ceiling of a room while screwing it into position is an example of static loading of muscle groups or static movement.

WORKPLACE LAYOUT AND DESIGN

Once the task analysis has been performed and ergonomic problems have been observed, the safety professional must determine how to eliminate the hazards. One method to achieve this goal is by examining how the work area is laid out and redesigning the workplace to eliminate the problems. Knowledge of workstations will help the safety professional in this process.

Classification of Workstations

There are several ways to classify workstations. One method of categorizing workstations is by tasks being performed. A second method is by using the worker posture requirements associated with the task. According to this second classification procedure, workstations are designed as seated workplaces, standing workplaces, or combination seated and standing. Ergonomists recommend seated workstations when

- All tools, equipment, or components required to perform the tasks are easily reached from the seated position.

- Tasks requiring fine manipulative hand movements or inspection activities are primarily being performed.

- Frequent over-head reaches are not required.

- Less than ten pounds of lifting force is required or the force exertion requirements to perform the task are minimal.

- There is adequate leg space available (adapted from Eastman Kodak, 1983).

Workspace design specifications include: a minimum of 20 inches in width, 26 inches in depth to allow for adequate leg clearance, a minimum of 4-inch clearance from the edge of the workstation, and an approximate ideal work area of 10-by-10 inches where tasks required activities are performed. Remember that seated workstations are recommended for detailed visual tasks, precision assembly work, or for typing and writing tasks.

Standing workstations are typically recommended when

- The operator is required to move around a given area as a requirement for performing the given task.
- Extended and frequent reaches are required to perform the intended tasks.
- Substantial downward forces or lifting heavy objects, typically weighing more than ten pounds are found.
- There is limited leg clearance that would permit a seated position below the workstation surface (adapted from Eastman Kodak, 1983).

Standing workstations are recommended for precision and detailed work, as well as for heavier work activities. Because of anthropometric differences in height for male versus female workers, the ideal working height for various tasks will vary. The anthropometric charts should be referred to for help in determining ideal measurements. Generating the maximum force in the upper body extremities requires that employees keep their elbows close to normal elbow heights standing or elbow elevations slightly below elbow height while standing. Have employees stand in a normal relaxed position and then bend their elbows at a 90-degree angle. The measurement between the elbow height and the floor will furnish the elbow height standing position. This measurement provides safety professionals with the ideal height for light-work-related activities while standing. Reduce the elbow standing height approximately four to twelve inches for heavier work requirements or raise the elbow height standing four to eight inches for more detailed work activities. Once again, standing workstations are preferred when precision work is required or when light and heavy assembly work is demanded. It is important for safety professionals to determine

- The tasks required to be performed
- The number of different employees that will share the workstation
- The degree of adjustability required to provide an optimal work area for all

The third category, the combination workstation, is recommended when

- Several tasks are performed by the employee where mobility is required
- Reaching over head or at levels below seated positions are required

This is especially true for reach activities on a variety of levels where forward or side position reaches must be accomplished at a variety of levels above the worksurface (adapted from Eastman Kodak, 1983).

Grandjean (1988) recommends seven guidelines for workplace layout and design:

1. Avoid any kind of bent or unnatural posture. (Bending the trunk or the head sideways is more harmful than bending forward.)
2. Avoid keeping an arm outstretched either forward or sideways. (Such postures lead to rapid fatigue and reduce precision.)
3. Work sitting down as much as possible. (Combination workstations are strongly recommended.)
4. Arm movements should be either in opposition to each other or symmetrical. (Moving one arm by itself sets up static loads on the trunk muscles. Symmetrical movements facilitates control.)
5. The working field (the object or table surface) should be at such a height that it is at the best distance from the eyes of the operator.

6. Hand grips, controls, tools, and materials should be arranged around the station facilitating the use of bent elbows close to the body.
7. Handwork can be raised up by using padded supports under the elbows, forearms, or hand.

ERGONOMIC HAZARDS

In this section, the ergonomic hazards associated with Manual Material Handling, Cumulative Trauma Disorders, and Video Display Terminal Workstation Design will be presented.

Manual Material Handling

From carrying luggage at the airport to moving boxes of supplies at the office, everyone carries and lifts objects. Moving objects, regardless of weight, can result in arm, back, and leg strain. In all occupational environments, manual material handling is a common activity. Workers handle raw materials, tools, finished materials, containers, and packing materials on a daily basis. This is an especially important problem in the health care industry as well. As a consequence of this daily exposure, material handling is the most serious cause of injury all over the world. The human suffering and costs associated with this problem are enormous. According to the National Safety Council's publication "Accident Facts," occupational injuries, as a whole, resulted in costs estimated to be in excess of $111 billion in 1993. Almost $31 billion or 28 percent of those total injury costs resulted from lifting problems. The losses that result from overexertion problems are compounded when factoring in the days absent from work. In 1993, approximately 28 percent of the total estimated days away from work in this country, or 21,000,000 days, resulted from lifting injuries. Over 500,000 workers in the U.S. experience lifting, lowering, pushing, pulling, and carrying injuries each year. Nursing and other health care staff are not immune to this problem. According to some recent estimates, more than 50 percent of the cost for all occupational injuries in health care could be attributed to back injuries resulting from patient lifting, transferring, and handling tasks (Watt, 1986/87). Because lifting is the most common injury associated with material handling, the safety professional should focus on how to prevent these injuries.

To gain an understanding of the causes of back injuries, the safety professional should refer to the 1991 NIOSH lifting guidelines. In this document, certain motions have been identified that increase the risk of back injuries. These movements include combinations of bending, twisting or turning, standing, and sudden position changes. NIOSH points out that a combination of bending and twisting/turning puts such a strain on the spine that it increases the likelihood for back injuries.

The primary hazard in lifting is based on the fact that the spine serves as a biomechanical lever. The force that results from lifting an object is multiplied by a factor of 10 because of the mechanical advantage produced by this biological lever. Twenty-four bones, called the vertebrae, compose the spine. A fluid-filled disc with a fibrous outer layer is located between each vertebra. These discs act as shock absorbers to prevent the rubbing and abrasion of the vertebra. Disc rupture or herniation are injuries that are often cumulative in nature. Typically, it is not the box that was lifted that caused the back pain, but a variety of other factors including age and repeated poor lifting practices that damaged the disk (Mital, et.al., 1993). When this type of injury occurs, the fluid in the disc pushes through the fibrous layer and applies pressure on the nerves in the spinal cord. Although disc damage is attributed to only about 5 percent of back injuries, it is often more serious and longer lasting than other forms of back injuries.

To avoid back injuries, employees must know proper lifting techniques. Traditionally, employees have been taught to use a 'bent knee/straight back' technique with the object between the knees. However, this can only be performed if the object fits between the knees. Research indicates that the legs may not always be strong enough to lift heavy objects using this technique, causing awkward and stressful postures (Garg and Herrin, 1979). There is no single lifting technique that applies to all situations, but the ideal goal for safety professionals is to eliminate manual material handling whenever possible. Consider engineering controls like cranes and conveyors to eliminate lifting and carrying activities. If engineering controls are not feasible, consider administrative controls such as assistance from co-workers.

Some additional rules of thumb that should be considered are

- Use engineering methods to automate lifting and moving.
- Ensure that the weight of the object is as close to the body as possible when employees must lift something manually.
- Eliminate bending and twisting motions when employees must lift objects.
- Require lifting aids, whether mechanical or co-workers, when exceeding NIOSH recommended weight limits.

CUMULATIVE TRAUMA DISORDERS

Cumulative Trauma Disorders (CTDs), also referred to as repetitive-motion injuries, are the result of excessive use of the hands, wrists, or forearm area. Like overexertion injuries, CTD frequency and costs are growing in epidemic proportions. The National Safety Council reports that 2,925,000 days or 3.9 percent of the total days lost from work, resulting from nonfatal occupational injuries, were due to repetitive-motion injuries. This ergonomic problem becomes a major concern to safety professionals when considering that 280,000 new cases of repetitive motion trauma occurred last year. This represents approximately 62 percent of all new occupational illnesses and about 4 percent of the total number of work related injury and illness according to the Bureau of Labor Statistics' estimates. Some of the most common cumulative trauma disorders are Carpal Tunnel Syndrome, Cubital Tunnel Syndrome, Tendonitis, and Tenosynovitis.

Carpal Tunnel Syndrome

Carpal tunnel syndrome is a common wrist injury caused by the compression of the median nerve in the carpal tunnel. The carpal tunnel is an opening in the wrist surrounded by the bones of the wrist and the transverse carpal ligament. The sensation in the thumb, index, and middle fingers is the responsibility of the median nerve. When the wrist is forced to flex or extend or deviate toward the ulnar (small finger) or radial (thumb) position, the ligament compresses the median nerve. In effect, these motions pinch this nerve as it passes through the carpal tunnel.

Symptoms that result from this pinching include tingling, pain, or numbness in the thumb and first three fingers.

The level of repetitive motion that can be tolerated without undue risk of injury varies widely depending on age, sex, and other health factors. While no specific limits have been established to cause CTD symptoms, factors that increase the risk include repetition of motion, work-rest cycles, force, and duration of the task. Task requirements should be modified if any of the following factors are observed:

- tasks requiring over 2,000 manipulations per hour
- manual task work cycles that are 30 seconds or less in duration
- repetitive tasks which exceed half of the workers' shift

In addition, safety professionals should be concerned when excessive force must be applied or if frequent use of stressful postures are necessary to complete a task. Medical conditions can also increase the risk of CTD. Studies indicate that arthritis, diabetes, poor renal, thyroid or cardiac function, hypertension, pregnancy, and fractures within the carpal tunnel increase the chances of an employee experiencing CTD symptoms (Parker and Imbus, 1992).

Carpal tunnel syndrome has been increasingly found in data entry clerks, cashiers, and individuals who perform keypunch tasks. This is due to the static, restricted posture and high-speed finger movement related to these jobs.

Cubital Tunnel Syndrome

Cubital Tunnel Syndrome is the compression of the ulnar nerve in the elbow. It is thought to be caused by resting the elbow on a hard surface or sharp edge. Its symptoms include tingling in the ring finger and little finger, the area where the ulnar nerve is responsible for sensation.

Tendinitis

Tendinitis is the most frequently diagnosed CTD. This disorder is the result of excessive inflammation of the tendon due to excessive use. The common symptoms are a burning sensation, pain, and swelling at various sites in the hand, wrist, or arm. Tendinitis is common for those workers required to extend their shoulders overhead during assembly activities.

Tenosynovitis

Tenosynovitis is the soft tissue trauma and injury to tendons or tendon sheaths. It frequently occurs in the wrists and ankles where tendons cross tight ligaments. The tendon sheath swells, making it more difficult for the tendon to move back and forth inside the sheaths. It often results from the over working of muscles or the wrenching and or stretching of tendons. Common forms of tenosynovitis are DeQuervain's disease and trigger finger (Parker and Imbus, 1992). These disorders are commonly found in employees whose jobs include buffing/grinding and packing because of repetitive wrist motions, vibration, and prolonged flexed shoulders. Tenosynovitis is often associated with ulnar deviation during rotational movements such as using a screwdriver.

It is important to recognize the symptoms of repetitive motions disorders. They have been described as appearing in three stages (Chaterjee, 1987). In stage one, victims may experience aches and tiredness; at this time, the disorder is fully reversible and may even subside after periods of rest. Stage two is characterized by the addition of swelling and pain. These symptoms do not dissipate overnight, and they usually last for several months and include a reduction in job performance. In stage three, victims may experience constant pain even when performing light duties. This condition can last for months or years. The key to an effective ergonomics program is to identify problems before they begin. Safety professionals should look for potential CTD problems during task analysis activities by using physiological motion categories that focus attention on awkward (unnatural) postures and frequent repetitions of motions.

Video Display Terminal Workstation Design

How many individuals work with computers on a regular basis? How many have more than one computer at home? The information age is here and video display terminal (VDT) workstations are common in homes, as well as in our offices and our automated production floors.

As technology and increased productivity have demanded workers to spend more time on computers, there has been an increased frequency of ergonomic injuries and illnesses in the workplace. Available research points to poor workstation design as the cause for both physiological and psychological problems for the employee (Danko, 1990). This has drawn attention to the need for an ergonomic study of computer or VDT workstations. The layout and design of a workstation must be geared to human capabilities and limitations in order to avoid physical discomforts, fatigue, mental stress, or injury to the worker (DeChiara, 1991). Computer workstations should be of great concern to safety professionals in both the layout of new facilities and improvement of existing office and other occupational settings (Danko, 1990).

Safety professionals must recognize that workstations are often used by several individuals of varying body size, weight, shape, age, physical condition, attitude, aptitude, and mental ability. Therefore, workstation adjustability is necessary to accommodate these factors. The work environment should incorporate anthropometric designs which enable workers to function within their capabilities. This allows all employees to perform their job tasks optimally and comfortably (Braganza, 1991). Anthropometric data supplies a standard guide of measurements, both static and dynamic, that provides an optimal dimensional layout of the workstation (Woodson, 1964).

In addition, other critical factors must be recognized during the workplace design or redesign process. These factors are: illumination, equipment and control display arrangements, warning signals, visual displays, control devices, traffic spaces, and storage requirements (Woodson, 1981). One important element to evaluate is the nature of the task. This evaluation will determine which type of workstation layout is appropriate for the setting. The three categories of workstations include seated workstations, (used for data entry and telecommunication jobs) standing workstations, (used for assembly and machine control jobs) and combination workstations (used for technical positions and computer design work). A wide variety of tasks are often performed in both office and manufacturing environments. As a consequence, a combination workstation often best meets the needs of these settings (Grandjean, 1988). The key factor to all workstation designs is having equipment and furnishings that are easily adjustable. This ensures that the equipment fits the individual properly, thus conforming to the individual's anthropometrical requirements. Workstations that require a seated worker should have easily adjustable chairs, worksurfaces, and equipment, as well as footrests for those requiring foot support to maintain their knees and legs on a horizontal level (Eastman Kodak, 1983).

A good chair is probably one of the most important workstation investments (Stewart, 1995). When a chair is properly adjusted, the worker should position her feet firmly on the ground. The hips and knees should assume a horizontal plane position at a 90-degree angle, and the pan of the seat should measure 15 to 20 inches from the floor. To further assume an ergonomically correct working position, workers' arms should dangle by their sides, then be raised to a horizontal plane position at right angles to their bodies. When the body is assuming these neutral positions, stress is relieved from the shoulders and back area, resulting in a decrease in fatigue and tension (Grandjean, 1988). To ensure that a chair has as many ergonomic comfort and safety qualities as possible, it should have the following features:

(1) adjustable backrest and seat (to fit different body sizes)
(2) good lumbar support (lower-back)
(3) five legs (for better stability)
(4) wheels (for mobility)
(5) reclining seat backs and adjustable seat pans (to control the pressure on the back and thighs)
(6) armrests and footrests(if necessary to the situation (Barnes, 1994)

Numbness, pain, and fatigue may still occur due to reduced blood circulation if an individual has been seated for an extended amount of time. Therefore, the furniture or equipment being used should have rounded edges to prevent discomfort (Braganza, 1994). In addition, frequent breaks and stretch activities should be incorporated into the employee's daily activities. Breaks do not necessarily imply coffee breaks. Changing in activities can meet the needs of an employee. Filing information while standing constitutes a break from computer data entry activities and will also provide therapeutic benefits.

The worksurface area, configuration, and height are some additional ergonomic concerns (Braganza, 1994) that safety professionals should not neglect. Pencils, paper, and other frequently used materials should be located near the body in a range of reach from 16 to 18 inches. Arm movements should also be made in a symmetrical or opposing action while the elbows are slightly bent. This prevents unnatural twisting or stretching of the body. Moving one arm by itself can cause static loads on the trunk muscles, whereas symmetrical movements allow for added control (Panero, 1979). Standard anthropometric data suggests that most desks are 30 inches high, which is too high for an ergonomically correct computer workstation. Depending on the height of the individual, the adjustability of the worksurface should be optimally between 23 and 28.5 inches high. This satisfies the seating requirements of 90 percent of seated adults (Barnes, 1994).

Thigh and knee clearance is another key consideration in designing a workstation (Barnes, 1994). A full range of movement and sufficient clearance for the thighs is necessary while sitting. Without sufficient clearance, an employee might twist the torso in order to get closer to the workstation. This twisting could cause injury. The minimum thigh clearance (measured from beneath the worksurface to the floor) is 20.4 inches (Braganza, 1994). If this clearance is not available, a standing workstation might be the layout of choice.

Anthropometric measurements should also be used to determine the proper angle and height of the monitor to reduce fatigue and eye strain. The top of the monitor should be aligned with the standard sight line, and the center of the screen should be 15 degrees below eye level (DeChiara, 1991). The screen should be tilted back zero to seven degrees to eliminate glare, and should be located between 18 and 28 inches from the eyes for ease in focusing (Grandjean, 1988). In addition, document holders should be placed at the same height and plane as the screen to decrease neck and eye fatigue (Barnes, 1994). The monitor, documents, and keyboard should be positioned to best suit the task being performed, reducing twisting and bending while using the equipment (Grandjean, 1988).

It is important to point out that purchasing ergonomically adjustable furniture will not necessarily eliminate ergonomic problems. Employees must know how to adjust the workstation to meet their specific needs. In several companies, it was determined that employees had ergonomically adjustable chairs. These workstations, however, were not properly adjusted because no one ever instructed these individuals that the chairs could be adjusted. In addition, correct ergonomic layouts were never explained to the employees. For ergonomic workstation programs to be effective, safety professionals must ensure that employees know ergonomic workstation layout and design, as well as how to adjust their furniture to eliminate ergonomic stresses. Training is an important part of all ergonomics programs.

CONCLUSION

Ergonomics has become an increasingly important part of the safety professional's job responsibilities. With an effective implementation plan, ergonomic activities can make a significant difference in the loss-control effectiveness of the organization. Through workplace layout and task analysis activities, the safety professional may identify many ergonomic hazards. Remember the operator-machine system, anthropometrics, and biomechanics during future walk-through inspections.

For some complex ergonomic hazards, organizational ergonomic factors must also be considered. Some of these hazards may be associated with factors such as shift work schedules or product flow problems. Work rest cycles may also be an issue. Depending on the nature of the work activities, psychological factors such as fatigue and boredom are other issues that can impact worker productivity and efficiency. No matter what the ergonomic problem might be, it is important for safety practitioners to use the operator-equipment system. This will ensure that all of the variables affecting ergonomic performance will be examined.

There are several hazards that are hot ergonomic issues today. Some of these include: manual material handling (Mital, et.al., 1993), cumulative trauma disorders (Putz-Anderson, 1988), manufacturing workstation layout and design (Eastman Kodak Company, 1983), video display terminal workstation design (Eastman Kodak Company, 1983), hand and power tools ergonomic issues (Parker and Imbus, 1992), occupational stress, shift work ergonomic hazards (Monk and Folkard, 1992) and environmental factors.

Questions

1. What are the components of the operator-machine system? Provide an example of a work activity using this system.

2. How can the use of the industrial hygiene triad be applied to ergonomic problems?

3. How can videotaping of jobs be useful in the evaluation of ergonomic problems?

4. How can Anthropometrics and Biomechanics be used to solve ergonomic problems?

5. How does Anthropometrics use statistical measurements and methods?

6. What are the costs associated with material handling and repetitive motion injuries? Why are these two types of ergonomic problems of concern to the safety professional?

References

Barnes, K. 1994, August. Is your office ergonomically correct? *HR Focus, 71,* 17.

Braganza, B. J. 1994, August. Ergonomics in the Office. *Professional Safety, 39,* 22-27.

Bureau of Labor Statistics. 1993. *Occupational injuries and illnesses in the United States by industry, 1991.* Washington, DC: U.S. Government Printing Office.

Chatterjee, D.S. 1978. Prevalence of vibration-induced white-finger in flourospar miners. *British Journal of Industrial Medicine*, 35:208-218.

Danko, S. 1990, Fall. Indoor ecology: Designing health and wellness into the workplace. *Forum, 19,* 3-8.

DeChiara, P. J., and Zelnik. M., 1991. *Time saver standards for interior design and space planning*. New York: McGraw-Hill.

DeVries-Griever, A.H. and Meijman, P.F. 1987. The impact of abnormal hours of work on various modes of information processing: A process model on human costs of performance. *Ergonomics*, 30: 1287-1299.

Eastman Kodak Company. 1983. *Ergonomic design for people at work*. Vol. 2. New York: Van Nostrand Reinhold.

Garg, A. and Herrin, G. 1979. Stoop or Squat? A biomechanical and metabolic evaluation. *Transactions of American Institute of Industrial Engineers*, 11:293-302.

Grandjean, E. 1988. *Fitting the task to the man* 4th ed. London: Taylor & Francis.

Kroemer, K. H., Kroemer, H. J., and Kroemer-Elbert, K. E. 1990. *Engineering physiology: Bases of human factors/ergonomics*. 2nd ed. New York: Van Nostrand Reinhold.

Laflin, K., and Aja, D. 1995. Health care concerns related to lifting: An inside look at intervention strategies. *American Journal of Occupational Therapy, 49*, 63-72.

Mital, A., Nicholson, A. S., and Ayoub, M. M. 1993. *A guide to manual materials handling*. London: Taylor & Francis.

Monk, T. And Folkard, S. 1992. *Making shiftwork tolerable*. London: Taylor & Francis.

National Safety Council. 1994. *Accident facts, 1994 edition*. Itasca, IL: National Safety Council.

Panero, J. and Zelnik, M. 1979. *Human dimensions and interior space*. New York: Watson-Guptill.

Parker, K. and Imbus, H. 1992. *Cumulative trauma disorders*. Boca Raton, FL: Lewis Publishers.

Putz-Anderson, V. 1988. *Cumulative trauma disorders: A manual for musculoskeletal diseases of the upper limbs.* London: Taylor & Francis.

Rutenfranz, J., Colquhoun, W.P., and Knauth, P. 1977. Biomedical and psychological aspects of shiftwork: A review. *Scandanavian Journal of Workplace Health*, 3:165-182.

Thompson, C. 1989. *Manual of Structural Kinesiology.* 11th ed. St. Louis, MO: Times Mirror/Mosby College Publishing.

Waters, T. R., Putz-Anderson, V., Garg, A., and Fine, L. J. 1993. The revised national institute for occupational safety & health (NIOSH) lifting equation, Bristol, PA: Taylor & Francis.

Watt, S. 1986/1987. Nurses' back injury: Reducing the strain. *Australian Nurses Journal, 16,* 46-48.

Woodson, W. E. 1981. *Human factors design handbook.* New York: McGraw-Hill Book Company.

Woodson, W. E. and Conover, D. W. 1964. *Human engineering guide for equipment designers* 2nd ed. Berkeley, CA: University of California Press.

Chapter 8

Fire Prevention and Protection

CHAPTER OBJECTIVES

After completing this chapter, you will be able to

- List the components of the fire tetrahedron.
- Define key fire terms.
- Identify the classes of fires and extinguishers.
- Explain the purpose of the NFPA and prominent safety-related regulations.
- Identify placards and labels.
- List the components of the fire program.

CASE STUDY

The Imperial Food Products plant, located in Hamlet, North Carolina, was owned and operated by Emmet J. Roe. Its 30,000 square-foot, one-story, brick-and-cinder- block facility had no windows but had nine exits, including the loading docks. The 200 employees were predominantly female. Chicken breasts were cooked and then frozen, packaged for resale to restaurant chains and frozen food companies. The facility did not have a building-wide sprinkler system at the time of the fire, but a carbon dioxide fire extinguishing system had been installed in the cooking area after a fire in the early 1980s. It was not known if the system was operational at the time of the fire (Rives, 1991, p. 1).

The Hamlet fire started when an overhead hydraulic line ruptured and sprayed flammable hydraulic fluid on the floor in the frying area. Natural gas burners, used to heat the large chicken fryers, ignited the vapors of the fluid and caused a fire that killed 25 workers and injured an additional 49 employees (Hughes, 1991, p. 8). As employees tried to exit the burning building, they found the fire exit doors locked. Management had locked several exits to prevent employees from stealing chickens. One door was kicked down as employees fled from the fire and suffocating smoke. A fire investigator found a door marked "Fire Exit-Do Not Block" padlocked. Next to the loading dock a door was locked. The main entrance door was also off its hinges. Some of the employees who could not find a way out tried to hide in the freezer where most of them died by suffocation. It was later learned that the employees of the Imperial Food Products plant had never received any type of fire evacuation training (Hughes, 1991, p. 8).

The Imperial Food Products plant opened its doors ten years prior to the fire. Over those ten years, the plant never had an inspection by the state, local, or federal inspectors. The town of Hamlet was not required to make inspections, although it is up to the municipality to enforce fire regulations. Ironically, every day the plant was in operation, there was a federal inspector on site to make sure that the chicken processed was acceptable for consumption. Locking a fire door is a violation of North Carolina's fire code, a misdemeanor in this case. The state fire code requires all municipalities to have a fire inspector even though a minimum number of inspections was not specified. Under the occupational safety regulations, a locked exit door would constitute a serious violation and because management locked these fire exits knowingly, it would be a willful violation (Diamond, 1991, p. 7).

One mandatory part of OSHA's General Industry Standards contains the Means of Egress regulations for evacuating workplaces during emergencies. The rule requires "a continuous and unobstructed way of exit travel from any point in a building" (OSHA's Standards for General Industry, 1993, p. 33). In 1980, OSHA added requirements for employee emergency plans and fire prevention plans to the "Means of Egress" regulations. Employers must develop emergency plans, create evacuation routes, install alarm systems, and create regular emergency response training programs. For the first time, victims of work related injuries and deaths could seek prosecution for the willful violations that caused bodily injuries (Diamond, 1991, p. 1).

Emmett Roe, the owner of Imperial Food Products, was sentenced to 19 years and 11 months in jail as part of a plea bargain that let his son Brad, the plant's operations manager, stay out of prison. The sentence was for illegally locking plant doors and not having a sprinkler system, leading to the death of the 25 workers.

The irony of this fire is that it is reminiscent of the infamous Triangle Shirtwaist Fire that occurred in New York City in 1911. One hundred and forty-six workers were killed and another seventy were seriously injured when fire broke out in a fireproof building where garments were manufactured. Some exits were only twenty inches wide and others were locked to prevent theft by employees. Once workers were able to break through a door to reach a fire escape, it soon collapsed. This fire is a strong indicator that the old saying "history repeats itself" is true.

FIRE TETRAHEDRON

Materials that burn are considered *combustible.* Those that burn at lower temperatures are considered flammable. Four components are necessary to sustain combustion: fuel, heat, oxygen, and a chemical reaction. A fire will not burn without the presence of all four. Fuels include numerous materials, both liquid and solid, but all must be in a vapor state before they can burn. Once vaporized, they mix with oxygen to form a combustible mixture which can burn when exposed to an appropriate temperature. During the burning process, a chemical reaction among the components of fuel, oxygen, and heat causes the chemical components to change form and release other gases. The following example will illustrate this process.

An owner of a boat sales and repair shop inspected a boat that his assistant was repairing in the shop. During the inspection, he smelled gasoline vapors, so he instructed his assistant not to activate a grinder until he had opened several doors to "air out the place". As he turned to enter the showroom, he heard the grinder turn on. He spun around to see about a dozen fires burning all over the repair shop. Both men escaped with minor burns, but the boat was destroyed, and the shop suffered extensive damage. The assistant assumed that there were not enough vapors to cause a fire. He may not have considered that the electrical grinder might spark when activated, and that it could ignite those vapors. Had the vapors not reached the lower flammable limit, they would not have ignited.

Lower flammable limit, also known as the *lower explosive limit,* is the lowest concentration of gas or vapor (percentage volume in air) that burns or explodes if an ignition source is present. The *upper flammable limit* or *upper explosive limit* is the highest concentration of a gas or vapor that burns or explodes if an ignition source is present. A mixture can have too little of the concentration (and be too lean) or too much of the concentration (and be too rich) to burn. In the case of the boat repair shop, the concentration was between the limits, so it was just right to ignite. Incidentally, the closer the mixture gets to the limits, the less complete the combustion. Obviously, a better mixture would be near the mid-point between the limits.

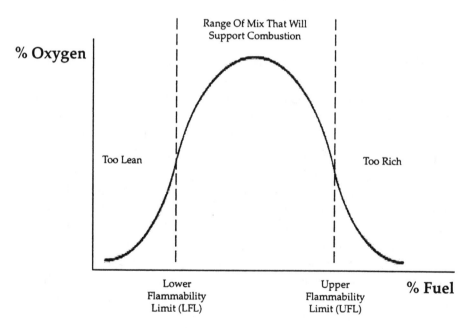

Optimum combustion occurs at the apex of the curve.

Figure 8-1. Flammability relative to the oxygen/fuel mixture.

The hazard with gasoline is that it has a low *flash point*. This is the temperature at which the substance gives off enough vapors to form a mixture which will ignite. Once exposed to a source of ignition, such substances will burn, or under certain circumstances, explode. The source of ignition might be a flame, a hot object, or even a spark from a tool or static electricity. Although it is possible for the mixture to exceed the upper flammable limit where it becomes too rich to burn, the safety practitioner should assume that any mixture above ten percent of its lower flammable limit poses a hazard and should be treated accordingly. All of these factors combined to cause an explosion when the assistant activated the grinder.

Another factor that becomes important is how certain items will burn. *Vapor density* becomes a key issue in determining where a vapor might be found in the atmosphere. If the vapor has a low density (below 1.0), it will float in the air. Therefore, ignition of such a fire could occur anywhere in a room or at a point above the source of the vapors. If the vapor has a high density (above 1.0), it will tend to move downward. Vapors released from a gasoline spill will often float downhill. An ignition source below the gasoline may ignite it and cause the fire to spread from the point of ignition to the source of the spill. This is one reason why police evacuate a wide area surrounding a gasoline spill on the highway. A nearby car could become a source of ignition and cause the remaining contents of the tank to burn and possibly explode.

The temperature at which a substance will burn is another way that materials are categorized. For example, liquids that have a flashpoint at or above 100 degrees (Fahrenheit) are called *combustible*. If the liquid ignites at a temperature below 100 degrees (Fahrenheit), it is referred to as *flammable*. The categories of combustible and flammable are subcategorized by the National Fire Protection Association (NFPA). Knowing these categories can be useful in determining how to protect some properties.

Figure 8-2. Example of fire extinguishers found in the occupational environment. (a) 150 pound wheel fire extinguisher for protection of facilities where larger fires could occur. (b) Hand held fire extinguishers. (Photographs provided by and reprinted with the permission of Ansul, Inc., Marinette, Wisconsin.)

Categories of Fires and Extinguishers

Fires are categorized according to types of materials involved:

- Class A fires involve ordinary combustible materials such as paper, wood, cloth, and some rubber and plastic materials.
- Class B fires involve flammable or combustible liquids, flammable gases, greases, and similar materials, and some rubber and plastic materials.
- Class C fires involve energized electrical equipment where safety requires the use of electrically nonconductive extinguishing media.
- Class D fires involve combustible metals such as magnesium, titanium, zirconium, sodium, lithium, and potassium.

Fire extinguishing agents are categorized by the types of fires they extinguish; that is Class A, Class B, Class C, or Class D extinguishers are used on corresponding types of fires. Some extinguishers can be used on different classes of fires; therefore, Class A-B and Class A-B-C extinguishers are available.

Each type can be recognized as follows. An extinguisher for Class A fires could be rated as 1-A, 2-A, 3-A, 4-A, 6-A, 10-A, 20-A, 30-A, and 40-A. A 4-A extinguisher will extinguish about twice as much fire as a 2-A extinguisher. Class B extinguishers are rated similarly. Class C extinguishers are tested only for electrical conductivity; however, no extinguisher gets a Class C rating without a Class A and/or Class B rating. Class D extinguishers are tested on metal fires. The agent used depends on the metal for which the extinguisher was designed. The extinguisher faceplate will indicate the effectiveness of the unit on specific metals.

A good example of a typical heavy-duty fire extinguisher can be found in Figure 8-2. Employees who use this type of extinguisher would typically be part of an on-site fire brigade.

National Fire Protection Association

The National Fire Protection Association (NFPA) was founded in 1896 to design sprinkler systems and develop techniques for installing and maintaining them. Since then, the organization has evolved to become the preeminent fire prevention group in the United States. Its goal is to safeguard people and their environment from destructive fires through the use of scientific and engineering techniques and education. The NFPA's activities involve the development, publication, and dissemination of codes and consensus standards (Ferry, 1990).

The NFPA's *Fire Protection Handbook* was first published in 1896 and is now in its 16th edition. The document is now 1,800 pages long and becomes increasingly larger over the years because of the expansion of the body of knowledge and the complexity of industries that have evolved (Ladwig, 1991). The NFPA also publishes other literature including: the *Life Safety Code Handbook,* the *Flammable and Combustible Liquids Code Handbook,* the *Hazardous Materials Response Handbook,* and the *National Electrical Code Handbook.* It publishes the *Industrial Fire Hazards Handbook,* which is not specifically a code handbook but is used as a fire-protection guide for industry. This book focuses on fire hazards and control methods that are associated with major industries and their processes (Ladwig, 1991).

Standards and Codes

Since the inception of the association, standards and codes have become more than a simple standardization, installation, and maintenance guide for sprinkler systems. There are currently about 260 codes that cover a large range of fire-related topics. There are approximately 200 NFPA committees that develop standards and codes using a democratic process. All are published in the *National Fire Codes,* or they can be requested individually in pamphlet form. The most widely used are NFPA 70, the *National Electric Code;* NFPA 101, the *Life Safety Code;* NFPA 30, the *Flammable and Combustible Liquids Code;* NFPA 13, the *Automatic Sprinkler Standard;* NFPA 58, the *Liquefied Petroleum Gases Standard;* and NFPA 99, the *Health Care Facilities Standard* (Ladwig, 1991).

NFPA 70

The purpose of NFPA 70 is to provide a guide for the practical safeguarding of persons and property from hazards arising from the use of electricity. It covers the installation of electrical conductors within or on public or private structures, the installation of conductors that connect to the supply of electricity, the installation of other outside conductors on the premises, and the installation of optical fiber cable (NFPA, 1988).

NFPA 101

NFPA 101 was designed to establish minimum requirements for life safety in buildings and structures. The *Life Safety Code* addresses a wide range of topics from fire and similar emergencies, construction, protection, and occupancy features necessary to minimize the dangers from fire, smoke, fumes or panic. It also identifies the minimum criteria for the design of egress facilities which permit the prompt escape of occupants from buildings, or where possible, into safe areas within the building. The *Life Safety Code* also recognizes that fixed locations which are occupied as buildings such as vehicles, vessels, or other mobile structures shall be treated as buildings. NFPA 101 does not attempt to address general fire prevention or building construction which is normally a function of fire prevention and building codes (NFPA, 1988); its only concern is the protection of occupants once a fire has started.

NFPA 30

The *Flammable and Combustible Liquids Code,* NFPA 30, was developed with the intent of reducing hazards to a degree consistent with reasonable public safety. It does not seek to eliminate all hazards associated with the use of flammable and combustible liquids because the goal was not to interfere with public convenience. This code applies to all flammable and combustible liquids except those that are solid at 100 degrees Fahrenheit (38.8 Celsius) or above. The code also sets the requirements for the safe storage and use of a great variety of flammables and liquids commonly available. NFPA 30 establishes requirements for the safe storage and use of liquids that have unusual burning characteristics or that are subject to self- ignition. This storage is also restricted based on heating conditions brought about by storage (NFPA, 1988).

An approved storage cabinet for keeping small containers of flammable and combustible liquids is shown in Figure 8-3. Approved containers are shown in Figure 8-4.

NFPA 13

NFPA 13 covers the installation of sprinklers by providing the minimum requirements for the design and installation of automatic sprinklers, including the character and adequacy of water supplies. In addition, it covers the selection of sprinklers, piping, valves, and all materials and accessories. It does not include the installation of private fire service mains, the installation of fire pumps, and the construction and installation of gravity and pressure tanks and towers (NFPA, 1988).

NFPA 58

NFPA 58 pertains to storage and handling of liquefied petroleum gases. The standard applies only to the highway transportation of LP-Gas and the design, construction, installation, and operation of LP-Gas systems. It does not include refrigerated storage systems, marine and pipeline terminals, natural gas processing plants, refineries, and petrochemical plants or tank farms (NFPA, 1988).

NFPA 99

The *Health Care Facilities Standard,* NFPA 99, created criteria to minimize the hazards of fire, explosion, and electricity in health-care facilities. These provide the basis for the performance, maintenance, testing, and safe practices for facilities, materials, equipment, and appliances and include other hazards associated with the primary hazard. NFPA 99 is intended for use by persons involved in the design, construction, inspection, and operation of health care facilities. It is also intended for the design, manufacture, and testing of appliances and equipment used in patient-care areas or health-care facilities (NFPA, 1988.)

Figure 8-3. (a) and (b) Flammable liquid storage cabinets with examples of different types of safety containers. (Photographs courtesy of Justrite Mfg. Co. L.L.C., Des Plaines, Illinois.)

Leaktight Closure. Spring-loaded, self-closing cap prevents spillage. Cork gasket resists virtually all chemicals; won't get brittle or crack for a safe, tight seal.

Ergonomic Handle Design. Swings free for easy carrying; compound lever action makes pouring easy.

Positive Pressure Relief. Automatically vents internal pressure at 5 psig to protect against explosion.

Welded Pad. Secures handle assembly to top.

Free-flow Flame Arrester. Double mesh, laminated screen dissipates heat to prevent flashback; removes without tools for cleaning.

Double-lock Seams. Four thicknesses of metal at breast and bottom seams, flooded with solder to prevent leaks. Every can is pressure-tested for leak tightness.

Durable, Powder Paint Finish. Maximizes chemical and abrasion resistance.

Yellow Wrap-around Label. Meets OSHA requirements for cans holding liquids with flash points of 80°F and lower.

Ribbed Bottom. Provides extra strength and rigidity. Raised to prevent accidental punctures

Heavy Gauge Terne-plate Construction Throughout.

FM Approved and UL Listed. Reliability backed by rigorous testing. Also meets OSHA and NFPA requirements.

Figure 8-4. Example of flammable liquid "safety can" with illustrated safety features. (Photograph courtesy of Justrite Mfg. Co. L.L.C., Des Plaines, Illinois.)

Educational Materials

Besides developing codes and standards, the NFPA also produces an abundance of educational materials for all age groups and levels of competence. They range from fire safety materials for children to fire ground tactics for professional fire fighters. These materials come in filmstrips, pamphlets, slides, movies, videos, posters, and books. There are educational materials for training the general employee population, the plant fire brigade, and for employee off-the-job fire safety education (Ashford, 1977).

NFPA 704

NFPA 704 is a system of marking storage containers to communicate the hazards associated with the contents of the container. NFPA 704 utilizes a diamond-shaped symbol broken into four quadrants at the top, bottom, left, and right of the diamond. The top portion is shaded red and represents the flammability hazard. Health hazards are indicated in the left portion which is blue. Yellow represents the corrosivity hazard. The bottom portion is white; any special hazards will appear here. For example, a W with a line drawn through it means the chemical is water reactive. A radioactive, poison, or other symbol might also appear here. In the top three blocks, the extent of each hazard is represented with a numbering system of 0 to 4. Zero represents a minimal hazard, while four represents the greatest hazard. These are demonstrated in the following chart.

SHIPPING PAPERS

The shipping paper represents the most vital piece of information available when you respond to a hazardous materials incident. The shipping paper contains information needed to identify the material(s) involved. Use this information to initiate protective actions for the safety of yourself and the public. The shipping paper contains the proper shipping name (see blue border index pages), the hazard class or division of the material(s), ID Number (see yellow border index pages), Packing Group, and, when applicable, Reportable Quantity notation (RQ) (for use in reporting spill incidents). In addition, there must be available information that describes the hazards of the material and that can be used in the mitigation of an incident. This must be entered on or be with the shipping paper. This requirement may be satisfied by attaching a guide page to the shipping paper, or by having the entire guidebook available for ready reference. Shipping papers are required for most hazardous materials in transportation. Shipping papers are kept in-

- the cab of the motor vehicle,
- the possession of the train crew member,
- a holder on the bridge of a vessel, or
- an aircraft pilot's possession.

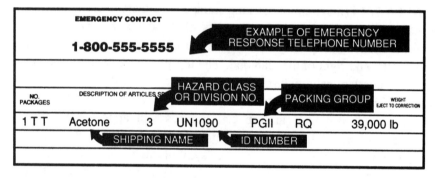

EXAMPLE OF PLACARD AND PANEL WITH ID NUMBER

The 4-digit ID Number may be shown on the diamond-shaped placard or on an adjacent orange panel displayed on the ends and sides of a cargo tank, vehicle, or rail car.

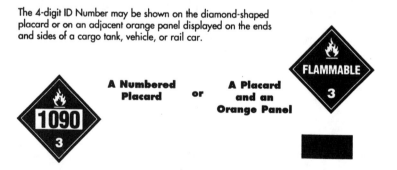

Figure 8-5. Shipping papers and example of placard and panel with ID number. (Courtesy of Emergency Response Guide, U.S. Department of Transportation.)

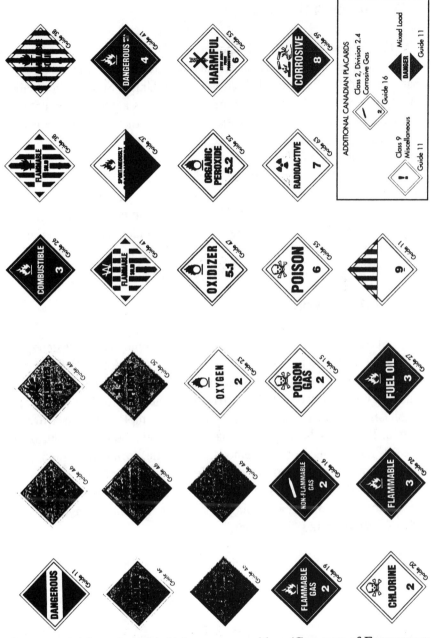

Figure 8-6. Placards and initial response guides. (Courtesy of Emergency Response Guide, U.S. Department of Transportation.)

DOT MARKING SYSTEM

The Department of Transportation (DOT) Hazardous Materials Regulations specify the requirements for placarding and labeling of hazardous materials being shipped within the United States. These labels are approximately 4-inches by 4-inches; placards are approximately 10-3/4 inches by 10-3/4 inches. Labels are attached directly to the container or on tags or cards attached to the containers. Placards are placed on both ends and each side of freight containers, cargo tanks, and portable tank containers. DOT utilizes United Nations' Hazard Class numbers that appear on the bottom of the label or placard, as well as a hazard-class, four-digit identification number. This number can be crossed referenced in the DOT *Emergency Response Guidebook* to identify the material and to learn of protective measures which should be taken to protect from a spill or leak.

OSHA REGULATIONS

OSHA deals with fire protection from an employee safety standpoint and many of the points that are covered in the OSHA standard are good management practices for property safety as well. Subpart E, Means of Egress, is taken from NFPA 101-1970, the *Life Safety Code*. The emphasis of this subpart is on protecting the employee once a fire has started. It informs the employer what to do protect the workers during the fire by addressing egress methods, automatic sprinkler systems, fire alarms, emergency action plans, and fire prevention plans.

Means of egress refers to a continuous and unobstructed way of exit travel from any point in a building or structure to a public way. It consists of three separate and distinct parts:

- the way of exit access (such as an aisle or hallway)
- the exit (such as a doorway)
- the way of exit discharge (such as a sidewalk outside the building)

Exits cannot be disguised or obstructed by mirrors, decorations, or other objects. They should also be marked by a readily visible sign. Any door, stairway, or other passage that might be confused as an exit should be marked "Not an Exit" or identified as Linen Closet, Basement, and so on, so that employees will be able to find their way out of the building.

The *Life Safety Code* of the National Fire Protection Association (NFPA) specifies the numbers and types of different occupancies and the sprinklers and alarms that are appropriate for each. Sprinklers and alarms are not required in all instances.

Subpart L of the OSHA regulations addresses fire protection. It contains requirements for fire brigades, all portable and fixed-fire suppression equipment, fire detection systems, and fire and employee alarm systems.

Safety practitioners have the responsibility to ensure that companies are in compliance with all applicable fire laws and that workers remain safe from fire hazards. This includes helping employees learn how to prevent fires and ensuring that workers are protected if a fire begins. Typical approaches to the fire problem begin with good *housekeeping*. The safety practitioner will monitor the conditions in the facility relative to neatness and cleanliness. Fires can begin as a result of poor housekeeping conditions such as a stack of oily or greasy rags being left in a pile or an open container. As the rags decompose, they become hot and sometimes ignite through a process known as *spontaneous combustion*. Other materials such as damp hay, grass, or paper products that have been exposed to dampness or a combustible liquid can undergo the same process. The fire begins spontaneously with no apparent outside ignition source. Other poor housekeeping conditions to be aware of include spilled lubricants or other combustible chemicals that have vapors which may come in contact with a source of ignition. Oily rags should be kept in approved "safety cans" as shown in Figure 8-7. to avoid contacts with ignition sources and incidents involving spontaneous combustion. Sometimes blocked means of egress can occur because packing materials, tools, or equipment that have been left in aisleways or in front of doors. It is the safety practitioner's job to monitor these hazardous conditions and ensure that they be corrected.

Figure 8-7. Example of oily waste "safety can" (Photograph courtesy of Justrite Mfg. Co. L.L.C., Des Plaines, Illinois.)

Safety practitioners will also take responsibility for assuring that fire alarm and suppression systems are in good working order. Requirements for this area are not only in fire and safety regulations, but are also requirements from insurance companies. The safety practitioner may work with the loss-control representative from the insurance company to ensure that all fire prevention and protection systems are in good working order.

Together, these professionals consider *alarm systems* and methods of alerting workers in the event of a fire or other emergency. These can include mechanical methods such as bells or sirens located throughout the plant that will be activated when an emergency occurs. Some small businesses may depend solely on someone yelling, "fire." They will also look at suppression systems such as fire extinguishers, sprinklers, or standpipes.

Fire extinguishers were discussed earlier. *Sprinklers* are the devices, often located on the ceiling, that release water when a fire occurs. Typically, they are activated by the melting of a fusible link when the fire gases get hot enough to melt that link. Different types of sprinklers include both *dry* and *wet* systems. *Dry systems* are often used where freezing conditions are found so the water will only be released into the pipes when a remote sensor activates the system. Once the temperature reaches a certain point, water is released into the pipes. In a *deluge* system, all sprinkler heads are already open. Some systems, however, will not release water until the sprinkler head itself is also activated. Sprinklers are an important resource in protecting lives and property. Properly installed and maintained sprinkler systems are nearly one-hundred-percent effective in protecting building occupants from fire.

Standpipes include the piping and hoses located throughout buildings that can be accessed during a fire. Some standpipes are *dry*, meaning that no water is in the pipes until pumped in during the emergency. In a tall office building, for example, the hoses and pipes in the standpipe system may be for use by the fire department. When firefighters arrive at the scene, they do not want to take the time to drag hoses to the top floors. A dry standpipe system could be in place which will permit them to connect to a water supply at the base of the building and activate the standpipe system on the top floors. Until they connect their water supply, however, no water is available. These systems often employ large diameter hoses that would be too unwieldy for untrained persons to handle.

Wet standpipes are typically designed for use by the occupants, if properly trained, to protect the people and the contents in the building. These systems have a constant supply of water in them and are smaller in diameter so they can be used by the occupants of the building. The loss-control representative of the insurance company will attempt to ensure, through the insurance contract, that all suppression systems are operating at all times, or that standby substitute arrangements are in place during periods of shutdown for maintenance and repair.

Some companies depend on *fire brigades* as a form of defense against fires. Fire brigades are units consisting of employees who have been trained to respond to fire emergencies. These units may have been taught to use only fire extinguishers, or they may have been prepared to a level comparable to that of members of a municipal fire department. Some companies maintain equipment including fire trucks and turnout gear like that found at the fire department. Unless employees have the appropriate training and equipment available to them to fight fires, it is recommended that employees be instructed to immediately evacuate and sound an alarm when a fire occurs. Even an *incipient* or early-stage fire can present a threat to the lives of employees when untrained personnel attempt to extinguish it. For liability and insurance purposes, many companies are reluctant to assume the risk of utilizing an in-house fire brigade as a defense for fires on their properties.

FIRE CASE HISTORY

This account was submitted by Greg Roberts, a master's degree student, who had worked for a department store chain.

In August of 1989, while safety manager, I received a call from Waldorf, Maryland, notifying me that a store had been partially destroyed by fire. The details were at best sketchy, so I boarded our corporate plane immediately to survey the damage and report on the cause.

The store was of a typical design with a 54,000-square-foot interior and an additional 5,000-square-foot garden shop attached outside. It met all building codes in Maryland for fire protection and was less than one year old.

When I arrived, I was met by the store manager and the inspector in charge of the investigation. What we discovered was that the fire began in the garden shop area and spread via the cantilever roof to the inside structure. However, the fuel source was unbelievable to the store manager and myself.

As many large retailers do, the buying department purchases boxcar loads of merchandise. The merchandise is then segregated and shipped in large quantities to each store. During the summer months, the store stocked and sold a great deal of peat moss, cow manure, and fertilizer of the 10-10-10 variety.

This merchandise was unloaded in the parking lot with forklifts either provided by the shipper or available at the store location. During the shipping process, unloading process, or storage process, the plastic bags containing the peat moss were damaged. This allowed rain water to accumulate inside the plastic bags while in our parking lot. The peat moss, usually in 40-pound bags with 40 to 50 bags per pallet, was located in the parking lot until it was needed in the garden shop area. When the restocking took place, an entire pallet (40 to 50 bags) was retrieved from the parking lot and placed under the cantilever roof for customer convenience.

The bags were protected from rain after being placed under the roof line, but they were not protected from direct sunlight during the afternoon portion of the day. This allowed the sun to heat this material and, as a result, ignite the peat moss, or as described by the inspector, allowed "spontaneous combustion" to occur. We had never experienced an event like this, and, fortunately, no one was injured in the fire. The vast majority of the merchandise was destroyed by the sprinkler system as the valves were chained in the open position, locked, and blocked by the last shipment of merchandise that was unloaded. These violations prevented the fire department from reaching the shutoff valve and, as a result ,the sprinklers operated well after the fire was extinguished.

We learned a great deal from our misfortune in Waldorf, Maryland. We were able to establish a storage policy that mandated that all spontaneously combustible material be kept either away from the building in the parking lot or maintained inside to reduce exposure to the elements. Additionally, we reinforced our policy regarding locked and blocked sprinkler valve shutoff and took disciplinary action against the store manager in Waldorf for these violations (personal communication, March 29, 1995).

MANAGING THE FIRE PROGRAM

Safety practitioners should carefully evaluate all in- and out-of-plant systems to assure that plant and emergency response personnel will be notified in a timely manner. If fire disables the alarm system or telephones, backup systems should be available and employees should know to utilize them. An effective fire plan should incorporate as many of the emergency systems into the daily plant operations as possible in order to assure that they are in operational order and that personnel are aware of how to use them.

Management commitment is critical to the success of the fire program and it can be measured in terms of the resources and the time that management commits to the program. There should be clear lines of authority that lead from the personnel responsible for assuring a successful fire program to the manager of the organization. A fire program is somewhat different from other safety programs in that equipment that is used in the fire program is spread throughout the plant. It may be impractical to have supervisors handle inspections in their respective sections and then provide a report back to the safety department. A better approach would be to have the safety practitioner work with a member of the maintenance or engineering staff to handle these responsibilities. A company that utilizes a fire brigade would have specialized personnel in charge of that group. It is good practice to involve as many different personnel in the fire program as possible to encourage support for the program.

WRITTEN PROGRAM

In 29 CFR 1910.38, OSHA requires companies to have an emergency action plan in writing that covers the following elements:

- Procedures for emergency escape

- Procedures to be followed by employees who remain to operate the plant
- Procedures to account for all evacuees
- Procedures for rescue and medical personnel
- Protocols for alarm systems
- Procedures for training

A written fire prevention plan is also required and should contain the following elements:

- A list of major workplace fire hazards and their proper handling and storage procedures, potential ignition sources and their control procedures, and equipment or systems which can control a fire involving them
- The names or job titles of personnel responsible for maintenance of equipment and systems installed to prevent or control ignitions or fires
- The names or regular job titles of the personnel responsible for control of fuel-source hazards

Training is critical. Employees need to be aware of the alarm systems that are used and what to do in the event that an alarm is sounded. In a part of the country where the possibility exists for different types of disasters to occur, employees need to be made aware of the different types of alarms. For example, a continuous bell may indicate that an evacuation of the building is necessary as would be the case with a fire. An intermittent bell may indicate that employees should seek appropriate shelter in the building as would be the case with a tornado.

Employees should also be trained on how and when to use fire extinguishers. Employees are only to use fire extinguishers if they have been properly trained; to do otherwise would be a violation of OSHA regulations. The reason for this is so that employees do not attempt to fight fires that are beyond their capabilities or the capabilities of the available equipment. If training is not mandated for employees, they have no alternative but to evacuate when a fire occurs.

CONCLUSION

Although most safety practitioners are not required to know the minute details regarding fire safety, it is imperative that they become aware of the terminology surrounding fire to be able to effectively communicate with building inspectors, fire marshals, OSHA regulators, fire equipment vendors and maintenance personnel, and insurance representatives. They must also maintain an awareness of the conditions that can lead to fire and work to eliminate them throughout the facility. An effective fire program requires constant vigilance and an ongoing partnership with key representatives from industry, government, and the insurance companies. The safety practitioner should cultivate relationships with as many of these representatives as possible to maintain an effective fire prevention and protection program.

Questions

1. What is the NFPA? Contact the NFPA and request a catalog of its products. The NFPA can be reached at (617) 770-3500, or you can write to them at One Batterymarch Park, Quincy, MA 02269. Explain what products are available from the NFPA and how these products would be of help to you in industrial safety.

2. Can you find any reference to NFPA in the OSHA Standards? Where? How are you affected by the NFPA and ANSI standards? How are they the same or different from laws?

3. Why do you think NFPA evolved as it did? Are the issues tackled by the NFPA issues that the federal government should be addressing? Explain your answer.

References

Ashford, N. 1977. *Crisis in the workplace.* Cambridge, MA: MIT Press.

Chissick and Derricott. 1981. *Occupational health and safety management.* New York, NY: Wiley and Sons.

Diamond, R. 1991, September 4. Plant never had safety inspection. *The Raleigh News and Observer,* pp. 1A, 7A.

Ferry, T. 1990. *Safety and health management planning.* New York, NY: Van Nostrand Reinhold.

Follmann, J. F. 1978. *The economics of industrial health: History, theory, practice.* New York, NY: AMACOM.

Hughes, J. T. 1991, September, 15. Nowhere to run. *The Raleigh News and Observer.* pp. 1J, 8J.

Lacayo, R. 1991, September 16. Death on the shop floor. *Time.* 28-29.

Ladwig, T. W. 1991. *Industrial fire prevention and protection*. New York: Van Nostrand Reinhold.

Meier, K. J. 1985. *Regulation: Politics, bureaucracy, and economics*. New York, NY: St. Martins Press.

National Fire Protection Association.(various). *National fire codes*. Quincy, MA: National Fire Protection Association.

National Safety Council. 1983. *Protecting workers lives: A safety guide for unions*. Washington, DC.: National Safety Council.

"Price of Neglect." September, 28, 1992. *Time*. 24.

Rives, J. P. and Mather, T. 1991, September, 4. 49 Injured as doors bar safety routes. *The Raleigh News and Observer*. pp. 1A, 7A.

Robertson, J. 1975. History and philosophy of fire prevention. *Introduction to fire prevention* p.7. Beverly Hills, CA: Glencoe Press.

Chapter 9

System Safety

CHAPTER OBJECTIVES

After completing this chapter, you will be able to

- Describe the different elements that compose system safety.
- Explain the development of system safety into a discipline.
- Describe the importance of system safety today.
- Identify the elements of the system life cycle.
- Explain how system safety program is managed.
- List the different tools and techniques in the analysis of system safety.

CASE STUDY

It was a cold January morning in Florida. During the night, icicles had formed on most of the structures at the launch pad for the Space Shuttle Challenger. As the astronauts left for their spacecraft that morning, they all gave the "thumbs up" signal, including Christa McAuliffe, the first teacher to go into in space. Everyone in NASA was excited about this launch because it was further proof of how safe and routine spaceflight had become. The countdown proceeded, the rockets fired, and Challenger lifted from the pad. Seventy-four seconds after launch, the euphoria vanished. As the two solid-rocket boosters careened away from the cloud that was supposed to contain the spacecraft, flight controllers received no more data. Challenger had exploded, and the seven individuals on board were killed.

The ten-member presidential committee that investigated the accident found that cold temperatures had caused the o-rings in the solid-rocket boosters to become less effective. These o-rings were made of a rubber-like caulking material and were used to prevent the hot gases from escaping the joints of the boosters. The cold caused the o-rings to shrink and become less elastic so, when Challenger throttled up to 110 percent of rated engine output, hot gases escaped. These gases acted like a blow torch and caused the external tank, filled with hydrogen, to explode. In addition, the committee found that during the assembly of the solid-rocket booster, the o-ring at that critical joint had been pinched, which further diminished its effectiveness.

The committee cited problems with NASA management and quality-control procedures. It took over two years of procedural and material redesign before the shuttle was allowed to fly. The problems still have not been completely solved, however, because in August, 1995, the shuttle fleet was once again grounded when hot gases were found to be escaping from the o-rings.

DEFINITIONS

To understand system safety, one must know the fundamentals that go into defining the discipline. A *system* can be defined as a group of interconnected elements that are united to form a single entity. Systems may include something as simple as a toaster or as complex as a chemical refinery. Perhaps the most important thing about systems is that they can sometimes and often are further defined into subsystems, assemblies, subassemblies, and components (*see* Figure 9-1). If the subsystems contain interdependent entities, they then also can be defined as a system. For example, a car brake can be defined as either a stand-alone system or as a subsystem of an automobile.

Safety can be defined as making something free from the likelihood of harm. Roland and Moriarty (1990) state that "safety in a system can be defined as a quality of a system that allows the system to function under predetermined conditions with an acceptable minimum of accidental loss" (Roland, 1990, p. 7).

A *hazard* is anything that can possibly cause danger or harm to equipment, personnel, property, or the environment. It is a circumstance which has the potential, under the right conditions, to become a loss.

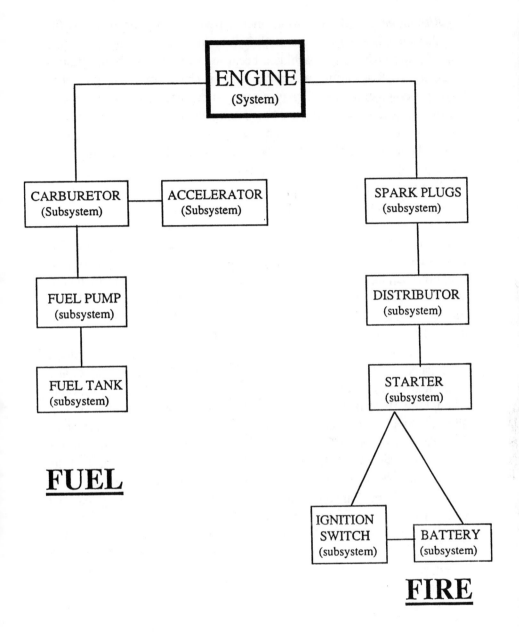

Figure 9-1. Block diagram showing the different parts of car engine (system)

Risk is the probability that an incident will occur. For example, the risk that the moon will hit the earth is very low. On the other hand, the risk of a driver having an accident because of insufficient brake fluid is much higher. Risk is often associated with the chance of occurrence. The batting average of a baseball player can also be described by a pitcher as the risk that the batter will get a hit. If someone has a batting average of 200, it would mean that if the pitcher threw ten pitches it would be likely that the batter would hit two fair. Obviously, the higher the risk, the more important it becomes to find and mitigate the hazard. In the case of the Challenger accident, the risk was high because failure of the o-rings would likely cause death if it occurred.

An *accident* is an unforeseen dynamic event that is caused by the activation of a hazard and consists of a number of interrelated events that result in a loss. It can cause the injury or death of individuals, as well as property damage to equipment and hardware. A related term that is sometimes used to refer to accidents but is actually a different event is an incident. An incident is also an unplanned event but may or may not have an adverse effect. An incident may simply be a mishap.

System safety can be defined as "an optimum degree of safety, established within the constraints of operational effectiveness, time, and cost, and other applications interfaces to safety, that is achievable throughout all phases of the system life cycle" (Malasky, 1982, p. 17). The system-safety concept deals with the before-the-fact identification of hazards as opposed to the after-the-fact approach that has been used for years. Consider the Challenger example; the accident occurred and the problem was found and resolved. Using the system-safety approach would mean that information concerning the performance of the o-ring material would have been gathered prior to design and testing and would have continued throughout implementation. There could have been a number of solutions to this problem. Even if no other material could have been found, procedures could have been in place that would have canceled the launch if temperatures were below a certain level. Assembly procedures could also have been modified so that misalignment of the solid-rocket sections and the potential pinching of the o-ring would be minimized.

HISTORY OF SYSTEM SAFETY

The first mention of the concept of system safety appeared in the technical paper *Engineering for Safety* presented at the Institute for Aeronautical Sciences in 1947. It stressed that safety should be designed into airplanes and continued by stating that safety groups should be an important part of the organization.

It wasn't until the early 1960s and the development of ballistic missile systems that the system-safety concept gained a more formal acceptance. At this time, contractors were given the responsibility for safety, replacing the practice of shared responsibility by each individual involved in the process. The first system safety requirements were published by the Air Force in 1962. This was modified in 1963 into the Air Force specification MIL-S-38130. In 1966, the Department of Defense adopted these specifications as MIL-S-381308A. MIL-STD-882 *System Safety Program for Systems and Associated Subsystems and Equipment; Requirements for*, developed in 1982, contained the specifications for the system-safety program required by all military contractors.

NASA also implemented a system-safety program patterned after the Air Force standards. These programs were instrumental to the successful completion of many NASA projects including the Apollo moon missions. The private sector has begun to develop system-safety programs because of the successes of the military and NASA. Leading the way are nuclear power, refining, and chemical industries. The adoption of system safety in those industries manufacturing consumer products has reaped the rewards in terms of more effective products, fewer accidents, and longer product life.

IMPORTANCE OF SYSTEM SAFETY TODAY

As society becomes more technically advanced, its tools become more and more sophisticated. In some cases, the machine has advanced further than the human capacity to control it. Jet fighters are good examples. These machines are capable of performing in g-forces that incapacitate most humans. Safety professionals need to be aware of the limits of human performance, as well as the fallibility of the individual in a mechanized system.

Product liability is also a major concern for many companies. The McDonald's coffee award is a case in point. In this case, a woman placed a cup of McDonald's coffee between her legs in a car. The coffee spilled and the woman was severely burned. McDonald's was found liable for the injuries, even though McDonald's cups contain a warning stating the contents are hot.

Table 9-1. System Life Cycle.

Phase	Safety Control Point	Result
Concept	Concept design review	Establish design for general evolution
Definition	Preliminary design review	Establish general design for specific development
Development	Critical design review	Approve specific design for production
Production	Final acceptance review	Approve product for release in deployment
Deployment	Audit of operation and maintenance	Control of safety operation and maintenance

SYSTEM LIFE CYCLE

The system life cycle consists of six phases: concept, definition, development, production, deployment, and disposition. At the end of each phase, a safety review is conducted. A decision is then made whether to continue the project or place it on hold, pending further examination.

During the *concept phase*, historical data and technical forecasts are developed as a base for a system hazard analysis. A Preliminary Hazard Analysis (PHA) is conducted during this phase. At the gross level, a Risk Analysis (RA) is performed to ascertain the need for hazard control and to develop system-safety criteria. Safety management will be doing the initial work on the System Safety Program Plan (SSPP). Three basic questions must be answered by the time the concept phase is completed:

1. "Have the hazards associated with the design concept been discovered and evaluated to establish hazard controls?

2. Have risk analyses been initiated to establish the means for hazard control?

3. Are initial safety design requirements established for the concept so that the next phase of system definition can be initiated?" (Roland, 1990, p. 23)

The *definition phase* is used to verify the preliminary design and product engineering. Reports presented at design review meetings should include the technological risks, costs, human engineering, operational and maintenance suitability, and safety aspects. In addition, subsystems, assemblies, and subassemblies of the system should be defined at this time. The PHA should be updated and a Subsystem Hazard Analysis (SSHA) should be initiated so it can later be integrated into the System Hazard Analysis (SHA). Safety analysis techniques should be used during this phase to identify safety equipment, specification of safety design requirements, initial development of safety test plans and requirements, and prototype testing to verify the type of design selected.

Environmental impact, integrated logistics support, producible engineering, and operational use studies are done during the *development phase*. The SSHA and safety design criteria should be completed during this time. During this period, interfaces with other engineering disciplines within the organization should be fostered. Using data collected during this phase, a go/no-go decision can be made before production begins.

The *production phase* of the system life cycle involves close monitoring by the safety department. In addition, the quality-control department becomes important because of its focus on inspection and testing of the new product. Training should also begin during this phase. Updating of the analyses started during the definition and development phases should continue. Finally, all the information collected during this phase should be compiled into the System Safety Engineering Report (SSER). The SSER identifies and documents the hazards of the final product or system.

When the system becomes operational, it is in the *deployment phase*. Data continues to be collected and training is conducted. If any problems occur, individuals responsible for system safety must be available to follow them up and decide on possible solutions. The system safety group in the organization should also review any design changes made on the system or product.

A sixth phase of the system life cycle, the *disposition* or *termination phase*, is the time that a product or system must be removed from service. In certain cases, the removal of a product may of itself create a hazardous situation. A good example would be when asbestos is removed from a building. Light transformer replacement may also create problems because of PCBs. Safety professionals should monitor these situations so both the worker and the public are protected.

MANAGEMENT OF SYSTEM SAFETY

Malasky (1982, p. 31) defines system safety management as

> ...that element of program management which ensures the accomplishment of the system safety tasks, including identification of system safety requirements; planning, organizing, and controlling those efforts which are directed toward achieving the safety goals; coordinating with other (system) program elements; and analyzing, reviewing, and evaluating the program to ensure effective and timely realization of the system safety objectives.

Using this definition, a block diagram can be drawn to illustrate the system safety function (*see* Figure 9-2).

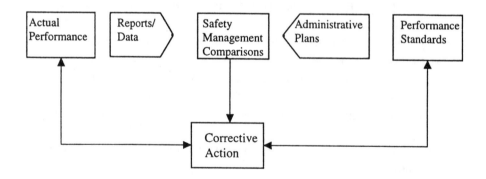

Figure 9-2. System safety functions.

Organizational Location

The system safety organization should have a position within the company where wide access to all areas of design, maintenance, and operation occurs. It can exist at many different levels, depending on the size of the organization. In large companies, the system-safety role may be found at all functional levels of the organization. In fact, MIL-STD 882B requires that all vendors submit a system-safety plan along with their proposals for federal contracts. Sometimes system-safety councils are used to reinforce corporate polices throughout the different levels of the company. Smaller companies may include the system safety work as part of each functional department. For example, the design department should have an individual whose responsibility would be in performing product or system safety analyses during the development phase of the system life cycle. However, it is up to each company to decide on which levels system safety should reside. The bottom line is to ensure that all risks have been determined and those that are unacceptable have been eliminated.

Organizational Interfaces

System safety analysis should not be performed in isolation. Input is necessary from nearly all functional disciplines within the organization. During the concept and definitions phases, system safety will primarily interface with the design group. Interfaces with engineering become important during the development phase. Quality assurance, training and development, industrial safety, and manufacturing engineering are important interfaces that should be developed during the production phase. Interfaces with maintenance and product-support disciplines are important during the deployment phase. The termination phase should find system safety interfacing with industrial safety, industrial hygiene, and product support.

Implementation

MIL-STD-882B (1984) has established the uniform requirement for implementing a system-safety program. Although this standard only applies to Department of Defense (DOD) "contract-definitized procurements; requests for proposals (RFP); statements of work (SOW); and Government in-house developments requiring system-safety programs for the development, production, and initial deployment of systems, facilities, and equipment" (p. 1), many companies use it to structure their system-safety effort. *Appendix A: Guidance for Implementation of System Safety Program Requirements* of the standard, provides a framework for achieving compliance.

Implementation Difficulties

Malasky (1982) discusses some of the difficulties which may be encountered during implementation. Problem formulation is cited as one issue. This may be caused by the conflicting demands of the various functional departments during the design optimization process or because of inadvertent minimization of potential hazards. For example, during the design process for the Ford Pinto, a fuel filler problem was discovered. Management decided that the cost of delay would be much higher than the cost of any lawsuits resulted from the hazard and went forward with production. In the long run, the cost to alleviate the hazard was much less than that of the lawsuits and loss of consumer confidence.

Organizational interfaces may pose other problems. When management does not take responsibility for decision making in the process, functional and system safety groups may be at odds with each other. Management perception is another difficulty. System-safety concepts are more abstract then many other disciplines. Therefore, it is important for the safety professional to structure the program so its impact is clear.

Elements of a System-Safety Program Plan (SSPP)

According to MIL-STD 882B (1984) the SSPP must specify the four elements of an effective system-safety program:

- A planned approach for task accomplishment
- Qualified people to accomplish the tasks
- Authority to implement the tasks through all levels of management

■ Appropriate resources for manning and funding to assure that tasks are completed

To accomplish these objectives, the SSPP should describe

1. The safety organization
2. System safety program milestones
3. General system safety requirements and criteria
4. Hazard analyses techniques and formats
5. System safety data
6. Safety verification
7. Audit programs
8. Training requirements
9. Mishap and hazardous malfunction analysis and reporting
10. System safety interfaces

For more detailed information regarding the specifics for each of these areas, the safety professional should refer to the standard.

TOOLS AND TECHNIQUES

In item four of the SSPP, a company must identify the types of techniques used in analyzing and evaluating system hazards. The following section will discuss some of the tools commonly used by the safety practitioner.

Preliminary Hazard Analysis

A Preliminary Hazard Analysis (PHA) is the initial effort in identifying hazards which singly or in combination could cause an accident or undesired event. PHA is a system-safety analysis tool used to identify hazard sources, conditions, and potential accidents (Roland & Moriarty, 1990). At the same time, PHA establishes the initial design and procedural safety requirements to eliminate or control these identified hazardous conditions. A PHA is performed in the early stages of the conceptual cycle of system development. It can be performed by engineers, contractors, production line supervisors, or safety professionals. Management should always first look at any risk involved in the operation of the system.

After identifying hazards and their resultant adverse effects, the analyst should recognize and rate each according to the Hazard Classification class which could be one of four categories:

- **Class I** - *Catastrophic:* A condition(s) that will cause equipment loss and /or death or multiple injuries to personnel

- **Class II** - *Critical:* A condition(s) that will cause severe injury to personnel and major damage to equipment, or will result in a hazard requiring immediate corrective action

- **Class III** - *Marginal:* A condition(s) that may cause minor injury to personnel and minor damage to equipment

- **Class IV** - *Negligible:* A condition(s) that will not result in injury to personnel, and will not result in any equipment damage

Roland and Moriarty (1990) show how to develop a Hazard Assessment Matrix to determine a Hazard Risk Index using frequency of occurrence and hazard category.(*See* Figure 9-3.)

Subsystem Hazard Analysis

A subsystem hazard analysis is performed to identify hazards in the component systems within a larger system. For example, in the Challenger accident, the solid-rocket boosters could be considered a subsystem. When the hot gases broke through the o-ring, a component of the subsystem, a total system breakdown began as a cascade effect which ultimately destroyed the orbiter. This analysis should be started no later than the definition phase in the system life cycle and should continue until the beginning of the system production phase. Analysis techniques include Fault Hazard Analysis (FHA) and Fault Tree Analysis (FTA) which will discussed in more detail in the next section (Roland, 1990).

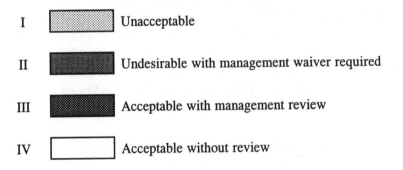

Frequency of Occurrence	Hazard Categories			
	I Catastrophic	II Critical	III Marginal	IV Negligible
(A) Frequent	1A	2A	3A	4A
(B) Probable	1B	2B	3B	4B
(C) Occasional	1C	2C	3C	4C
(D) Remote	1D	2D	3D	4D
(E) Improbable	1E	2E	3E	4E

The Hazard Risk Index (HRI) is then determined by reading the chart and assigning risk into one of the four levels of acceptability:

I ▢ Unacceptable

II ▢ Undesirable with management waiver required

III ▢ Acceptable with management review

IV ▢ Acceptable without review

Figure 9-3. Hazard Assessment Matrix. (Adapted from Roland (1990))

Hazard Analysis Techniques

The role of the safety professional is to anticipate, identify, and evaluate hazards; give advice on the avoidance, elimination, or control of hazards; and attain a state for which the risks are judged to be acceptable. To achieve this, they adopt a system-safety concept that includes:

- an understanding of the hazards,
- an understanding of the risks,
- an identification of the hazards and risks within their system,

■ an understanding of unwanted releases of energy and unwanted releases of hazardous materials being the causal factors for hazard related incidents, and

■ a knowledge of the principles and techniques used to control hazards and their associated risks to an acceptable level (Manuel, 1993).

A *hazard analysis* is used to identify any dangers that might be present in a proposed operation, the types and degrees of accidents that might result from the hazards, and the measures that can be taken to avoid or minimize accidents or their consequences (Hammer, 1989).

System Hazard Analysis (SHA) is primarily performed during the definition and development phases of the system life cycle (Roland, 1990). However, it should be continuously implemented throughout the life cycle of a system, project, program, and activity to identify and control hazards. The purpose of performing an analysis during the early stages of life cycle is to reduce costs. If the analysis is done after the system is in operation, then the system may need to be redesigned and consequently withdrawn from service. In addition, if the system is close to the end of its life cycle, it might not be cost-effective to change it (Brauer, 1990).

A *hazard analysis* should contain the following information:

■ Descriptive information

  -System mode
  -Subsystem mode of subsystem of hazard origin
  -Hazard description
  -Hazard effects
  -Likelihood or relative likelihood of each hazard

■ Causation events of each hazard

  -Identify events precisely as to subsystem mode, system mode, and environmental constraints

- Subsystem interface problems of special significance

  -Identify subsystem involved
  -Identify system and subsystem modes

- System risk evaluation

  -Severity listing of each hazard
  -Likelihood of each
  -If full quantitative evaluation is conducted, it should be presented for each hazard

- Risk summary

  -Listing risks of each hazard and for the system as a function of system modes
  -A logical evaluation of acceptability of system risks
  -Recommendations as to system risk control

This analysis can and must begin as soon as the idea for a new system or operation is conceived (Roland, 1990).

Technic of Operations Review (TOR)

In 1987, D. A. Weaver (Ferry, 1988) developed the *Technic of Operations Review (TOR)*. It was designed to uncover management oversights and omissions instead of hardware or operator problems. The four steps of a TOR analysis are *state, trace, eliminate,* and *seek*.

During the *state* portion of the analysis, detailed information about the hazard should be gathered. If the hazard is discovered as the result of an accident, a summary of the mishap report should be read and fully understood.

The *trace* portion uses a sheet which displays possible operational errors under eight categories. These categories are

1. **Training.** This includes all errors related to lack of or inadequate preparation of the employee.

2. **Responsibility.** This area looks at errors in the organizational requirements which may have contributed to the hazard.
3. **Decision and direction.** This looks at the lack of or inadequacy in the decision-making process which cause errors in the performance of the product or employee.
4. **Supervision.** Those errors that are due to problems with the direction of employee work.
5. **Work groups.** Those problems which can be traced to the interpersonal relationships within the group.
6. **Control.** This area deals with errors caused by the lack of or inadequate safety precautions.
7. **Personal traits.** Errors that can be traced to the individual's personality and how it affects the individual's job performance.
8. **Management.** Those errors that can be traced to poor managerial control.

Under each category is a list of numbered operational errors that are in that group. A complete copy of the TOR analysis materials can be obtained from The FPE Group, 3687 Mt. Diablo Road, Lafayette, CA 94549, who are the sole distributors of this analysis instrument. To begin the analysis, a single number or prime error under one of the eight categories is chosen. Using the Challenger mishap, the prime number may be under category five, control; number 61, unsafe condition. Each subcategory number on the TOR sheet then points to other possible factors that may have contributed to the hazard or mishap. Possible factors that could have contributed in the Challenger accident include:

65. Deficient inspection, report, or maintenance (main category: control)

36. Failure to investigate and to apply the lessons of similar mishaps (main category: decision & direction)

43. Unsafe act; failure to correct before the accident occurred (main category: supervision)

86. Accountability; failure to develop appraisal and measurement of key goals and objectives (main category: management)

These factors are then discussed by the group. Those found to have contributed to the accident can then be further broken down until all the causative factors have been identified.

When the trace portion of the analysis is finished, the *eliminate* step begins. Sometimes the trace can identify a large number of contributing factors, a number that is often overwhelming to the group evaluating the hazard. Therefore, the eliminate step is used to discuss each contributing factor and evaluate its merit to the process.

The final step, *seek*, looks for possible actions that need to be taken to correct the problem. These solutions should then be implemented.

Technique for Human Error Rate Prediction (THERP)

Swain and Rook developed Technique for Human Error Rate Prediction (THERP) in 1961 (Malasky, 1982). It is used to quantify human error rates due to problems of equipment unreliability, operational procedures, and any other system characteristic which could influence human behavior. THERP is an iterative process that consists of five steps which can be executed in any order:

- Determine which system failure modes are to be evaluated.
- Identify the significant human operations that are required and their relationships to system operation and system output.
- Estimate error rates for each human operation or group of operations that are pertinent to the evaluation.
- Determine the effect of human error on the system and its outputs
- Modify system inputs or the character of the system itself so as to reduce the failure rate (Malasky, 1982, p. 257).

THERP is usually modeled using a probability tree. Each branch represents a task analysis that shows the flow of task behaviors and other associations. A probability is assigned based on the event occurrence or nonoccurrence.

Failure Mode and Effects Analysis (FMEA)

A Failure Mode and Effects Analysis (FMEA) is a sequential analysis and evaluation of the kinds of failures that could happen and their likely effects, expressed in terms of maximum potential loss. The technique is used as a predictive model and would form part of an overall risk assessment study. This analysis is described completely in the MIL-STD-1629A. The FMEA is most useful in system hazard analysis for highlighting critical components (Ridley, 1994).

In this method of analysis, the constituent major assemblies of the product to be analyzed are listed. Next, each assembly is then broken down into subassemblies and their components. Each component is then studied to determine how it could malfunction and cause downstream effects. Effects might result on other components, and then on higher-level subassemblies, assemblies, and the entire product. Failure rates for each item are determined and listed. The calculations are used to determine how long a piece of hardware is expected to operate for a specific length of time. It is the best and principal means of determining where components and designs must be improved to increase the operational life of a product. In addition, it is best used to analyze how often and when parts must be replaced if a failure, possibly related to safety, must be avoided (Hammer, 1989).

FMEA is limited to determination of all causes and effects, hazardous or not. Furthermore, the FMEA does very little to analyze problems which could arise from operator errors or hazardous characteristics of equipment created by bad design or adverse environments. However, the FMEA is excellent for determining optimum points for improving and controlling product quality (Hammer, 1989). An example of a FMEA for a bicycle can be found in Figure 9-4.

Fault Hazard Analysis (FHA)

A *Fault Hazard Analysis (FHA)* is an inductive method for finding dangers that result from single-fault events (Roland, 1990). It identifies hazards during the design phase. The investigator takes a detailed look at the proposal system to determine hazard modes, what will cause the hazards, and the adverse effects associated with the hazards.

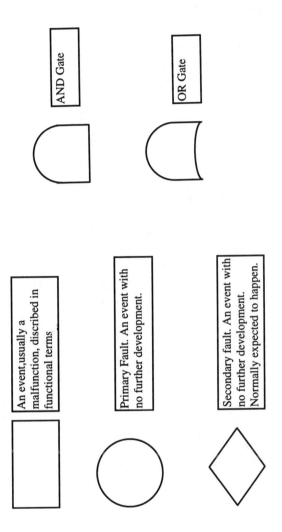

Figure 9-4. Basic FTA symbology

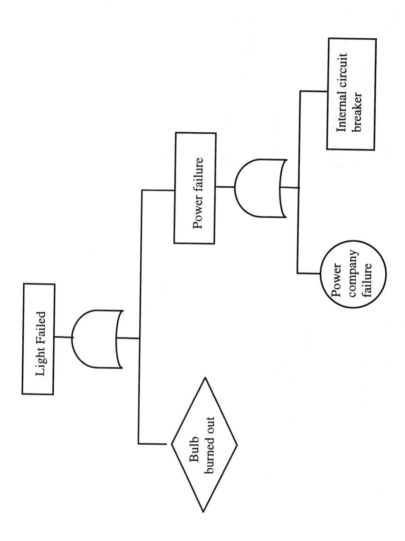

Figure 9-5. A simple FTA.

Fault Tree Analysis (FTA)

The most widely used analytical technique (Ferry, 1988), Fault Tree Analysis (FTA), is a symbolic logic diagram which graphically shows the cause-and-effect relationships of a system. A FTA uses the a basic symbology that includes:

The hierarchy of fault tree events can be classified as follows:

1. Head event (top event). The event at the top of the tree that is to be analyzed.
2. Primary event. The main malfunction of the component.
3. Secondary event. The effect that is caused on another component, device, or outside condition.
4. Basic event. "An event that occurs at the element level and refers to the smallest subdivision of the analysis of the system." (p. 153).

Using this basic hierarchy, an FTA is conducted using three steps. The first would be the determination of the head event. Once the head event is set, primary and secondary events can be determined. The final step is the resolution of the relationship between the casual events and the head event using the terms AND and OR.

An example of a simple FTA may help explain how AND and OR logic is used to define relationships. If an analysis is done on the failure of an electric light bulb, the following fault tree can be developed.

The tree shows two OR gates. After the light fails, there are two possible alternatives. Either the bulb will burn out *or* there is a power failure. If there is a power failure, there are two possible outcomes, either the power company has shut down *or* an internal circuit breaker has failed.

On the other hand, if there is an AND gate, both outcomes must occur for the event to happen. If an AND gate was placed after the power failure, both a power company failure and an internal circuit failure must happen for the event to take place.

There are limitations to this type of analysis. A specific knowledge of the design, construction, and the use of the product is necessary for successful completion of the FTA. In addition, FTA logic assumes that all events are either successes or failures. Many times such a concrete decision is not possible.

JOB SAFETY ANALYSIS (JSA)

Reasons For Conducting a JSA

Job Safety Analysis (JSA) is an analytical tool that can improve a company's overall performance by identifying and correcting undesirable events that could result in accidents, illnesses, injuries, and reduced quality and production. It is an employer/employee participation program in which job activities are observed; divided into individual steps; discussed; and recorded with the intent to identify, eliminate, or control undesirable events.

JSA effectively accomplishes this goal because it operates at a very basic level. It reviews each job and breaks it down into an orderly series of smaller tasks. After these tasks have been determined, the same routine of observation, discussion, and recording is repeated, this time focusing on events which could have a negative impact on each step in the task. Once potential undesirable events are recognized, the process is repeated for a third time and corrective actions are identified.

Conducting a JSA can be a valuable learning experience for both new and experienced employees. Not only does it help them understand their jobs better, but it also familiarizes them with potential hazards and involves them in developing accident prevention procedures. Workers are more likely to follow procedures if they have a voice in planning. Finally, the JSA process causes employees to think about safety and how it relates to their jobs.

Who Should Conduct JSAs

The responsibility for the development of a JSA lies with first-line supervision. These individuals have first-hand knowledge of the process, its potential hazards, and the need for corrective actions instituted at each step. This also provides the interaction with hourly employees necessary to complete the JSA. Initially, first-line supervisors must receive training in hazard recognition and procedures necessary to perform a JSA. This training will give them the knowledge necessary to explain the JSA to employees, what it is expected to accomplish, how it is conducted, and what their part will be in the program.

It has been proven that a well-organized and maintained JSA program can have a very beneficial effect on accident prevention, improved production, and product quality. Emphasis for this program, as with any other program, must start at the top and be conveyed down the line to all employees.

Procedures and Various Methods Used to Perform JSAs

A job safety analysis is a procedure used to review job methods and uncover hazards that

1. may have been overlooked in the layout of the plant or building and in the design of the machinery, equipment, tools, workstations, and processes;
2. may have developed after production started; or
3. may have resulted from changes in work procedures or personnel.

The principal benefits of a JSA include:

- giving individual training in safe, efficient procedures;
- making employee safety contracts;
- instructing the new person on the job;
- preparing for planned safety observation;
- giving pre-job instruction on irregular jobs;
- reviewing job procedures after accidents occur; and
- studying jobs for work-methods improvements.

A JSA can be performed using three basic steps, but a careful selection of the job to be analyzed is an important preliminary step.

Various Methods for Performing JSAs

There are three basic methods for conducting JSAs. The direct observation method (which will be discussed in detail in the next section) uses observational interviews to determine the job steps and hazards encountered. A second way to perform a JSA is using the discussion method. This method is typically used for jobs or tasks that are performed infrequently. It involves pulling together individuals who have done the job and having them brainstorm regarding the steps and hazards. The third way to perform a JSA is called the recall-and-check method. This method is typically used when a process is ongoing and people can't get together or to the worksite. Everyone participating in this process writes down his ideas about the steps and hazards involved in the job. Information from these individuals are compiled and a composite list is sent to each participant. Each person can then revise the list until consensus is achieved.

The following list gives the three basic approaches that can be used to determine how to perform a specific JSA.

- By a specific machine or piece of equipment (for example: lathe)
- By a specific type of job (for example: machining)
- By a specific occupation (for example: machinist)

Selecting the Job

A job is a sequence of separate steps or activities that together accomplish a work goal. Jobs suitable for a JSA are those which a line supervisor chooses. Jobs should not be selected at random. Those with the work accident history should be analyzed first if the JSA is to yield the quickest possible results.

In selecting jobs to be analyzed and establishing the order of analysis, top supervision should be guided by the following factors:

- Frequency of accidents. A job that has repeatedly produced accidents is a candidate for a JSA. The greater the number of accidents associated with the job, the greater its priority claim for a JSA

- Rate of disabling injuries. Every job having a history of disabling injuries should have a JSA performed. Subsequent injuries prove that preventive action taken prior to their occurrence was not successful

- Severity potential. Some jobs may not have a history of accidents but may have the potential for causing severe injuries. The more severe the injury, the higher the priority for a JSA.

- New jobs. Changes in equipment or in processes obviously have no history of accidents, but their accident potential may not be understood. A JSA should be conducted for each new job. Analysis should not be delayed until an accident or near miss occurs.

After the job has been selected, the three basic steps in conducting a JSA are

1. Breaking the job down into its component steps

2. Identifying the hazards and potential accidents

3. Developing solutions

1. Breaking the job down into its component steps

Before the search for hazards can be started, a job should be broken down into a sequence of steps, each describing what is to be done. There are two common errors in the process which should be avoided. They are

1. making the job breakdown too detailed so that an unnecessarily large number of steps results, and

2. making the job breakdown so general that the basic steps are not recorded.

To perform a job breakdown, use the following steps. (*See* Figure 9-6)

- Select the right worker to observe. Select an experienced, capable, and cooperative person who is willing to share ideas.
- Observe the employee performing the job.
- Completely describe each step. Each step should tell what is done, not how it is done.
- Number the job steps consecutively.

- Watch the operator perform the job a number of times until you are sure that all the steps have been noted.
- Check the list of steps with the person observed to obtain agreement on how the job is performed and the sequence of the steps.

2. Identifying hazards and potential accidents

The purpose of a JSA is to identify all hazards, both those produced by the environment and those connected with the job procedure. Each step must be made safer and more efficient.

Close observation and knowledge of the particular job are required for the JSA to be effective. The job observation should be repeated until all hazards and potential accidents have been identified.

The following list of hazard types will make it easier for the observer to make sure nothing was overlooked.

3. Developing solutions

The final step in a JSA is to develop a safe job procedure to prevent the occurrence of accidents. The principal types of solutions are

- find a new way of doing the job,
- change the physical conditions that create the hazards,
- change the work procedure, and/or
- reduce the frequency of the job.

Figure 9-6. Job Safety Analysis Worksheet

Job Name _____   JSA Number _____

Employee Name _____   Area/Supervisor _____

Employee Title _____   Last Analysis Date _____

Analysis By _____   Analysis Date _____

Job Steps	Potential Hazards	Necessary Safety Procedures	Required Safety Equipment

Completing the JSA

After the worksheet is completed, the data compiled should be transferred to the actual JSA form. A copy of a computerized template of a JSA can be found in Appendix A. Once the data has been entered and verified, it is important to obtain signature approval for the JSA from an upper-level manager.

Now that the JSA has been completed, it should be discussed with those employees performing that job. Any necessary safety procedures or additional safety equipment required to perform the job should be reviewed with these employees. In addition, a copy of the JSA should be available for employees to use when they perform the job. This is particularly important for those jobs that may not be done on a regular basis

Finally, it is important to note that no job is static. JSAs should be reviewed on a regular basis and any necessary changes should be made.

Effectively Using a JSA in Loss Prevention

The major benefits of a JSA come after its completion. Supervisors can learn more about the jobs they supervise. Employees who use JSAs have improved safety attitudes and their safety knowledge is increased. Supervisors can also use JSAs for training new employees. JSAs provide a list of the steps needed to perform the job, as well as identifying the procedures and equipment needed to do the job safely.

The JSA can furnish material for planned safety reviews. All the steps in the JSA should be followed with an emphasis on the major safety hazards. Supervisors should occasionally observe employees as they perform the jobs for which the analysis has been developed. If any procedural deviations are observed, the supervisor should alert the employee and review the job operation with her.

JSAs should be reevaluated following any accident. If the accident was a result of an employee failing to follow the JSA procedures, the fact should be discussed with all those who do the job. It should be made clear that the accident would not have occurred had the JSA procedures been followed. If the JSA is revised following an accident, these revisions should be brought to the employees' attention *(see* Figure 9-6 for a sample JSA worksheet).

An Example

Performing JSAs is a complex process. In order to better understand the importance of this analysis tool, Figure 9-7 shows a completed JSA for hydraulic line replacement. Please note accident categories described in the potential hazards section. These potential hazards will aid in determining what safety measures and personal protective equipment is necessary for a given job.

Accident Investigation

Accident investigation techniques have already been discussed in Chapter 3. However, it is important to note that accident investigation is a form of system safety analysis. During the deployment phase, accidents may occur and the findings of the investigation could result in a redesign, retrofit, and/or recall of the product or system.

Accident Types

The following discussion of accident types should assist you in determining the potential hazards for the job safety analysis.

1. **Struck-by.** A person is forcefully struck by an object.

2. **Struck-against**. A person forcefully strikes an object.

3. **Contact-by.** Contact by a substance or material that is by its very nature harmful and causes injury.

4. **Contact-with.** A person comes in contact with a harmful material.

5. **Caught-on.** A person or part of the person's clothing or equipment is caught on an object that is either moving or stationary.

6. **Caught-in.** A person or part of the person is trapped, stuck, or otherwise caught in an opening or enclosure.

7. **Caught-between.** A person is crushed, pinched, or caught between either a moving object and a stationary object or between two moving objects.

8. **Foot-level-fall.** A person slips, trips, and/or falls to the surface the person is standing or walking on.

---

9. **Fall-to-below.** A person slips, trips, and/or falls to a level below the one the person was walking or standing on.

10. **Overexertion.** The person performs a task beyond his physical capabilities, resulting in sprain or strain injuries.

11. **Exposure.** Employee injury results from her close proximity to harmful environmental conditions in the workplace.

CONCLUSION

System safety should be an important part of the product life cycle. The days of allowing the individual to find the problems is no longer acceptable. A proactive approach is necessary because of increased product-liability litigation and increased employee awareness of process-safety concerns. Therefore, the safety professional should make every effort to understand the concepts of system safety.

JSA #: 1430	TASK NAME: HYDRAULIC LINES - CHANGING		PAGE: 1 OF 3
JOB TITLE: HYDRAULIC LINE		AREA: MAINTENANCE	
MILL:	REVISION DATE: 08/15/94	WRITTEN BY: JOE HAMMERSMITH	
INTERVIEWS WITH: ED SMITH		APPROVED BY: DAVID JONES	
1994 REVIEWED BY:	1995 REVIEWED BY:	1996 REVIEWED BY:	1997 REVIEWED BY:
1998 REVIEWED BY:	1999 REVIEWED BY:	2000 REVIEWED BY:	2001 REVIEWED BY:

TASK STEPS	POTENTIAL HAZARDS	SAFETY PROCEDURE TO FOLLOW	SAFETY EQUIPMENT
1) Put on all required safety equipment, then report to job location.	*Contact By:* Forklifts/Blind Corners	When going anywhere in the mill, always walk to one side of pathways. Look before rounding corners.	Hard hat Safety glasses Steel-toed boots Hearing Protection Gloves Lock-out tags
	Foot-Level Fall: Oil, water, or trash on floor.	Avoid wet, oily spots. Pick up or step over trash or bunks. If your shoes/boots get slippery, clean them.	
2) Lock-out the system for the hydraulic hose.	*Caught in:* You may become entangled in the machinery.	**All workers must place their Lock-out Tag on all system breakers.**	Lock-out tags
3) Bleed off system pressure if needed.	*Foot-Level Fall:* The floor is slippery from hydraulic fuel.	Walk around spilled fluid if possible. Use Oil Dri to soak up spills.	Oil Dri
4) Remove the fitting on end of hose.	*Contact By:* Hydraulic fluid will irritate and damage the eyes and skin. *Contact with:* Tools may slip and cause you to hurt your hands.	Loosen the fitting to bleed off any remaining pressure in the hose. Keep the hose away from your face and others. Wear safety glasses. Use the right size tool on the coupling. Keep tools clean from oil and grease.	Safety glasses
5) Repeat step #4 on the other hose end.	See Step #4.	See Step #4.	See Step #4

Figure 9-7. JSA for a hydraulic line replacement.

JSA #: 1430	TASK NAME: HYDRAULIC LINES - CHANGING		PAGE: 2 OF 3
TASK STEPS	**POTENTIAL HAZARDS**	**SAFETY PROCEDURE TO FOLLOW**	**SAFETY EQUIPMENT**
6) Cut a new hose the same size as the old hose.	*Contact By:* Your hand may get cut by the saw.	Keep your hands away from the saw blade while it is turning.	
	Contact By: Flying sparks and debris can hurt you.	Wear safety glasses. Hold the hose with one hand while cutting it.	Safety glasses
7) Press the hose end and the Coupling Press.	*Caught In:* Your hand can be crushed if the press starts.	Move hand away from the ram and hose, then start the press.	
8) Lower the pressing collar onto the coupling.	*Caught In:* Your hand can be crushed if the press starts.	Make sure the press is off prior to inserting the hose and coupling.	Hard hat Safety glasses Steel-toed boots Earplugs Lock-out tags
9) Repeat Step #7 and Step #8 for the other hose end.	See Step #7.	See Step #7.	See Step #7.
10) Install the new hose onto the machinery.	*Caught In:* You may become entangled in the machinery.	**Make sure that the system is Locked-out and properly tagged.**	Lock-out tags
	Contact By: Any sharp edges of the coupling will cut your hands.	Make sure there are no burrs on the couplings. If so, grind them off.	
	Contact With: Tools may slip and cause you to hurt your hands.	Use the right tool size on the coupling. Keep tools clean from oil and grease.	

Figure 9-7. *(cont'd)*

JSA #: 1430	TASK NAME: HYDRAULIC LINES - CHANGING		PAGE: 3 OF 3
TASK STEPS	**POTENTIAL HAZARDS**	**SAFETY PROCEDURE TO FOLLOW**	**SAFETY EQUIPMENT**
11) Remove all Lock-out Tags from the system breakers	*Caught In:* Anyone still working on the hose may become entangled in the machinery.	Make sure all workers and tools are clear of the machinery. **Do not remove another person's Lock-out Tag**	
12) Check for any hydraulic oil leaks.	*Contact By:* Hydraulic fluid will irritate and damage the eyes and skin.	Stand away from hose while system pressure increases. Wear safety glasses.	Safety glasses
	Foot-Level Fall: The floor is slippery from hydraulic fluid.	Walk around spilled fluid if possible. Use Oil Dri to soak up spills.	Oil Dri
13) Tighten any leaky couplings.	See Step #10.	See Step #10.	See Step #10
14) Return equipment to the golf cart.	*Contact By:* Dropping tools may hurt your feet.	Carry all tools with a firm grip. If necessary, make several trips to carry all tools. Wear steel-toed boots.	Steel-toed boots
	Contact With: Various hazards in the mill.	Be alert for low-hanging items and moving machinery/ vehicles.	Steel-toed boots

Figure 9-7. *(cont'd)*

Questions

1. What steps make up the product life cycle? Include examples of the activities that occur during each step.

2. What is the importance of the system safety function to an organization?

3. What are the differences between system safety and industrial safety?

4. Can you develop a Fault Tree for a simple system? Make sure to include both AND and OR gates.

5. Using the hazard assessment matrix, what is the level of risk for the following activities?

   - Getting hit while walking across a busy street
   - Having a grease fire in the kitchen
   - Being involved in a plane crash

6. Can you perform a job hazard analysis of a simple system using established criteria?

References

Brauer, R.L. 1993. *Safety and health for engineers*. New York. NY: Van Nostrand Reinhold.

Department of Defense. 1984. *MIL-STD-882B: Task section 100: Program management and control*. Washington, D.C: U.S. Government Printing Office.

Ferry, T. S. 1988. *Modern accident investigation and analysis* (2nd ed.). New York, NY: John Wiley & Sons.

Hammer, W. 1989. *Occupational safety management and engineering*. Englewood Cliffs, NJ: Prentice Hall.

Laing, P. M., Ed. 1992. *Accident prevention manual for business and industry: Engineering and technology*. Itasca, IL: National Safety Council.

Malasky, S. W. 1982. *System safety: Technology and application*. 2nd ed. New York, NY: Garland STPM Press.

Mansdorf, S.Z. 1993. *Complete manual of industrial safety*. Englewood Cliffs, NJ: Prentice Hall.

Manuel, F.A.1993. *On the practice of safety*. New York, NY.: Van Nostrand Reinhold.

Ridley, J. 1994. *Safety at work*. Oxford, England: Butterworth-Heinemann Ltd.

Roland, H.E., and Moriarty, B. 1990. *System safety engineering and management*. 2nd ed. New York, NY: John Wiley & Sons.

Chapter 10

Managing the Safety Function

CHAPTER OBJECTIVES

After completing this chapter, you should be able to

- Explain the functions of management and how they relate to the job of safety.
- Identify the purpose of line and staff and the difference between the two.
- Differentiate between audits and inspections.
- Identify the role of safety in the staffing process.
- Explain the OSHA guidelines for safety management.

CASE STUDY

Bob Renee was hired fresh out of college by a small company to "do safety." Because Bob had a degree in occupational safety, the owners believed that he would be in a position to handle the safety function for a small facility. Bob felt confident that he understood the technical aspects of safety; he even had a handle on management. What he really did not understand, however, was how it all fit together. As a result, Bob left after a few months because he "needed more direction from his superiors."

Safety, like many other management activities, consists of planning, organizing, controlling, and directing, and, possibly, staffing. Although some may argue that once the organization is staffed, this function may cease, in most companies, staffing is an ongoing problem and its importance is paramount to the success of the enterprise.

PLANNING

"A job well planned is a job well done." "A job well planned is a job half finished." These axioms have been repeated in management classes for decades, and they hold some truth. Planning is essential to the ongoing success of any enterprise and certainly to the components of the enterprise, including safety. A well-planned operation involves a series of deliberate steps. First, the safety practitioner must forecast the needs of the safety department for the coming year. This involves reviewing the records of successes and failures as well as all of the resources used in the past. It also means anticipating the obstacles that may be encountered during the coming planning period. Most companies tend to operate on a one- to five-year cycle with planning for budgets occurring each year. This *forecasting* of coming needs or predicting when they will occur is a result of looking at the past and studying the future. There are a number of texts that can guide the reader as to how to forecast.

Once forecasts are made, the practitioner must then anticipate the resources needed to meet those demands and make requests accordingly. A proposed safety standard that will be coming into effect may require unusual demands as will the planning or start-up of a new facility. The safety practitioner must forecast these needs before they occur. This is part of a *proactive* as opposed to a *reactive* approach to safety. The practitioner does not wait for incidents to occur, but rather anticipates and *plans* to deal with problems before hand.

Established plans often become standards by which the practitioner can judge the performance of the safety program. They should evolve from the mission and the objectives of the organization. This can occur by carefully reviewing organizational goals with top management and writing safety objectives that complement or aid the organizational goals. Once safety objectives are written and methods by which those objectives can be accomplished are laid out, then budgets and timetables are formulated. A careful review of the documents associated with the planning tools is made with management and agreement is sought. Once this occurs, the organization of resources to accomplish the objectives can begin.

Astute safety practitioners soon realize that management supports their efforts to the extent that these efforts support those of the organization. Resources necessary to accomplish the task will be allocated on a basis relative to how well the accomplishment of the task helps the organization achieve its goals. Of course, personalities, personal ambitions, and politics can interfere in the process. Generally speaking, though, it is critical for efforts of the safety department to be tied to those of the organization.

ORGANIZING

Unlike most management activities, safety usually operates from a purely staff position. The implications of this pervade safety and affect the way it operates in the organization. To understand the management of safety, an understanding of line and staff positions is essential.

Line positions are charged with carrying out the major function or functions of the organization. In an army, the line fights; in a production operation, the line does the producing; in a sales organization, the line sells. First-line supervisors, plant managers, and even company presidents are considered line officers within an organization. *Staff positions*, on the other hand, are charged with supporting or helping the line positions. Typically, staff positions have no real authority over line activities. Staff members only assist and advise the line officers. Any authority which staff positions are given beyond that of assisting or advising is only a result of a line manager giving that authority. In some organizations, staff members are given authority over certain functions and any activities that relate to those functions. If a safety member in such an organization recognizes a problem in safety, he or she could be given authority to correct that problem. Even in organizations where no *functional* authority is given, some members of line management may perceive that the authority exists anyway. This could be particularly true of lower members in line positions who perceive that staff members have authority over them, when, in fact, they do not.

Safety managers, safety engineers, safety professionals, safety technicians, or whatever their titles are nearly always considered staff personnel and nearly always operate from a staff position. Their job is to monitor safety, compare what they find against existing standards, and advise line management as to any corrective actions that need to be taken from a safety standpoint. Their ideal position within the organization is reporting to the chief executive officer, but, typically, the safety practitioner reports to personnel or human resources departments. The rationale is obvious. No person within the organization is exempt from safety. If safety is subordinate to personnel, human resources or some position other than the chief executive officer, the opportunity for abuse and the potential for conflict of interest is high. Proper monitoring and appropriate advising of the enterprise cannot take place from the middle of the organizational hierarchy.

The full-time safety practitioner usually has no direct line authority over line managers. This person monitors safety and advises management on what changes are needed and why. To do otherwise would be to interfere with the jobs of the line managers. If a conflict occurs between the safety practitioner and the production manager, the production manager's orders will take precedence. Exceptions to this rule sometimes occur when the staff person is given *functional authority* or authority over the safety function. He or she has authority over all matters as they relate to safety and can overrule the line manager if his orders contradict those of the safety practitioner relative to safety. Of course, the safety practitioner may also have line authority over any personnel that work for and report to him or her such as an industrial hygienist, an occupational nurse, or a safety assistant.

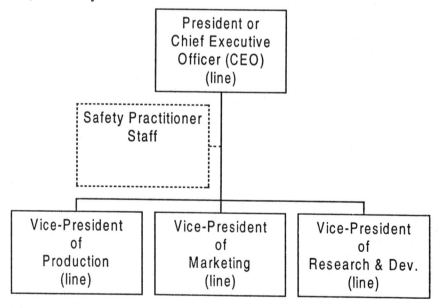

Figure 10-1. Typical organizational chart.

Generally speaking, the safety practitioner will be in a better position to positively influence the organization if a purely advisory role is maintained. It permits her to help line managers perform by becoming part of their team and also helps avoid conflicts. This approach squarely places the responsibility for safety on the shoulders of line management. If an unsafe condition occurs, the line manager cannot reasonably state that he was unaware of the condition. It's his job to be aware of unsafe conditions and have them corrected.

All members of the organization must buy in to safety if a safety culture is to be established. This will only happen with the support and continual input from top management. The safety practitioner does not lead the safety effort but rather acts as a member of the different safety teams throughout the enterprise. Teams may be in the form of safety committees, individual crews, or work cells. It is the safety practitioner's responsibility to help those teams help themselves relative to safety. She may act as a resource or even a guide in helping the team to establish itself. Management of safety is then up to the leader of the work group and the individual members.

Carl Wagner is supervisor of an eight-member production group. The group's job is to produce a high-quality, pearl-like finish on metal furniture. Carl's responsibilities are to see to it that quality production occurs *safely*. When the plant safety manager observes one of Carl's workers not using the prescribed personal protective equipment, she addresses the problem with Carl. If Carl is not around, then the safety manager will discuss the problem with the worker. It's actually Carl's job to see that the employees in his crew work safely. If Carl learns that the employee is not wearing the personal protective equipment because it is uncomfortable, then he will discuss the issue with the safety manager. The safety manager's job is to aid Carl and to help him resolve the employee's problem. Because of this relationship, the safety manager is not considered a threat to Carl, but a resource he can call on to help him and his crew to do a better job.

On Monday morning, Carl approaches Janet Reeves, the safety manager, and complains that the new respiratory equipment is too hot, and his employees do not like to wear it while working with lubricants. If Carl believes that the safety equipment is simply a nuisance that hinders his employees and tries to obstruct the safe practice methods used in the plant, it will be Janet's job to convince Carl that the equipment is necessary to the safety of his workers and to help him convince his employees.

Janet discusses the problem with Carl, and they decide to take time during the next "tool box talk" to talk it over with employees. Before going into the meeting, Janet makes Carl aware that OSHA requires that respiratory equipment be worn in this operation and that working without it could cause the employees to develop respiratory problems. After a long discussion, Carl decides to lead the talk himself and to ask his employees if they understand the consequences of not wearing the equipment. After a little verbal sparring, the employees reluctantly agree when they realize the importance of wearing the equipment.

Janet also suggests that it be possible to shorten the time that employees wear the respirators. After some discussion, all agree that any job demanding more than thirty minutes should be performed by two employees. After one employee wears the respirator for thirty minutes, he can switch off to permit another employee to finish. No employee will have to wear a respirator for more than thirty minutes at a time. All agree that this is a workable idea and they decide to give it a try.

If Carl is unable or unwilling to comply with his safety responsibilities, Janet is in a position to discuss the matter with Carl's boss, his boss's boss, or even the chief executive officer of the company. With top management support of the safety effort, Carl's performance will be measured on how well he supervises *safe* production. If he is unable or unwilling to produce safely, then his performance ratings will suffer. Therefore, he will be penalized as he would for failing to follow other company procedures regarding production or any other critical activity within the company.

CONTROLLING

Controlling occurs through a number of subfunctions. It involves looking at what is happening in the organization by monitoring, comparing the results of the observations to established standards, and then taking appropriate corrective actions. This occurs through inspections, audits, records reviews, interviews with employees and supervisors, and a careful watch on what is happening in the organization.

The results of monitoring are compared to results from previous years, existing safety regulations, and published or internally developed standards. Any deficiencies are noted and plans are made for correction. Before drastic changes occur, management approval and support is sought. Once this happens, it is up to line management to then make any and all appropriate corrections. Of course, management will only make corrections when it perceives that they are appropriate and beneficial. This perception is frequently based on line management's understanding of the need and the consequences of not making corrections.

The safety practitioner often finds that being a persuasive salesperson is as important as knowing what to do and what to sell. The safety practitioner soon learns that requests for change typically cost money, time, or other resources. In addition, these requests are competing with those from production, marketing, and other branches of the organization whose managers believe that they can best utilize the company resources.

It becomes apparent that requests by the safety function for company resources are competing with every other worthy cause within the organization. If money is spent on safety, it may not be available for raises for personnel, a new dental benefits plan, research on a breakthrough for a promising new drug, or a marketing effort that could reap millions. The management team will consider the safety request, weigh it against those from the rest of the organization, and respond accordingly.

If the safety professional is unable to make a convincing case for project resources, then the project won't happen. Management support for safety is a result of its perception of how well safety supports the organization. If the safety department is well aligned with the mission of the organization and the safety professional can build a strong case, usually in terms of the cost benefit to the organization, then safety will be supported. However, if management does not perceive that safety supports the organization or does not provide an adequate return, then it will not support it. To reiterate, the role of the safety professional is to monitor what is occurring in the organization relative to safety, compare what is found against existing standards, and advise on changes that need to be made.

Allegedly, the major tool used by professionals to monitor the state of safety in the organization has been the *audit*. Audit is a term loosely used by professional and semi-professional safety practitioners. It can mean anything from a cursory inspection of hand tools by shop personnel to a complete review of the safety program by the safety staff and numerous collaborating personnel. In reality, the audit is a tool that permits the assignment of a quantitative or numerical value to some aspect of the safety program. It is used to determine where that program is relative to where it ought to be. As many as a dozen or more audits might be done on an annual basis in any company. An example will illustrate this point.

Suzy Harris is the safety manager for Atlantic Widgets. She has developed nine safety programs in the following areas:

> Hazard Recognition and Control
> Fire Prevention and Protection
> Industrial Hygiene Monitoring and Control
> Ergonomics
> Waste Prevention and Minimization
> Security
> Manpower Training and Development
> Environmental Operations and Emergency Response
> Technical Standards and Legislative Compliance

Suzy remains open to new ideas for safety programs and plans to initiate more as time permits. Each of the above programs is reviewed annually in a different month by other plant managers and members of the safety committee. Suzy always tries to bring in at least one outside person to get fresh insight into what she is doing.

In January, she audits the Fire Prevention and Protection program. During the audit, Suzy attempts to determine whether the company is in a state of readiness for any fire eventuality. She does so by evaluating the systems that are in place to assure continuity of an effective program through interviews, observations, and a thorough records review.

Suzy and two members of her team interview Steve English, the plant manager, and ask him, "How are your supervisors specifically evaluated, for purposes of promotion or salary increases in the performance of their jobs, relative to fire prevention and protection?" She will assign a subjective rating to the following table based upon her evaluation of Steve's response:

All plant supervisory personnel are evaluated on their performance in the area of fire prevention and protection for purposes of promotion and salary increase.

No evaluation exists	Evaluation occurs but there is a link to salary promotion.	An effective evaluation occurs; there is a marginal impact on salary and promotion	An effective evaluation occurs and appropriate impact on salary and promotion
1	2	3	4

Suzy continues with similar questions developed to measure the program. She may have developed these questions with the help of others in the plant, or she may have purchased an audit instrument from a firm specializing in safety consulting. After assigning scores to all of the questions, Suzy totals the scores and makes a determination, based on a predetermined scale, as strength or weakness of the fire prevention and protection program.

As a side benefit, Suzy is also in a position to make recommendations based on her findings as to what changes need to be made to strengthen the program. These can be prioritized by what the cost of not making the changes might be. The following paragraph provides an example.

Suzy learns in her audit that the only protection for an infrequently used warehouse is provided by fire extinguishers. The contents of the warehouse are considered to be museum pieces and the owners are reluctant to part with them. Suzy makes a determination of the probability of a fire occurring in the warehouse in any given year. She can do this based on the past record of a large company or by estimates provided by her insurance company. The calculations indicate that a fire would likely occur only once in every 200 years with projected losses expected to run about $400,000. This amount assumes that the nearby fire department would be able to extinguish the fire with only fifty percent losses.

Next, Suzy uses the expected value technique and multiplies the probability of the loss occurring in a given year, 1/200 or .005 times the expected loss of $400,000 to estimate the expected value of a loss at $2000. This value approximates how much the company might be willing to invest in protecting the building from a fire in a given year. Two thousand dollars would be the break-even point in a given year. In other words, the company could only expect to save $2,000 in a given year by protecting the warehouse from fire.

How much would Suzy be willing to invest to save that money? It depends on a number of factors but primarily on how much return Suzy or the company can expect to get on its investments elsewhere. Suzy may decide that there are higher priorities for safety when she determines other findings through using this mathematical process. Even if she decides that this is a high priority, she must still convince management that it's the best investment it can make with the available funds. Whether she makes that recommendation depends on the return that management normally gets on its investments. If management expects a fifteen percent return in a given year, then Suzy will use the fifteen percent figure as her standard of performance. She won't make a recommendation that will not result in at least a fifteen percent expected return. If Suzy finds that she can come up with a relatively inexpensive system, she will make the recommendation based on her comparison to the standard.

Safety Audit Steps

1. Determine the area to be audited.

2. Develop or acquire an audit instrument based on the needs of the company. Be aware that no generic audit that is purchased or acquired from an outside firm will exactly fit the needs of the company.

3. Set up a pre-audit conference with all collaborating personnel. Discuss the purpose and scope of the audit.

4. Perform the audit, based on a predetermined evaluation system. Use interviews, observations, and records reviews.

5. Compare the audit results to the standards of performance established by the organization and the safety department.

6. Report to management based on the existing standards and variances from them. Suggest corrective actions and completion dates. Strong rationales must be prepared for any changes to which management might object. Changes must be tied to the organizational mission and objectives.

7. Follow-up to see that the changes agreed upon have been made.

8. Suggest corrective actions and negotiate completion dates with management.

After Suzy has been at the facility for a while, plant personnel begin to understand the expectations that management has for them regarding safety. Suzy finds it is only necessary to perform the audit once per year. Obviously, this is not frequently enough to monitor the overall fire program. Suzy has taught the supervisors to use their own employees to make periodic fire inspections. *Inspections* involve the monthly check of pressurized extinguishers, as well as comparing other safety features in the plant to given standards. The difference between audits and inspections is that in inspections, no quantitative evaluations are made. Suzy simply asks, "Is this done or not?" The inspection determines adherence to the operating standards. Audits are used to find out if inspections are taking place as well as the effectiveness of those and other safety tools.

Suzy also makes periodic inspections simply to ascertain that the supervisors are doing their jobs and to assure compliance with the key safety standards of the organization. The supervisors realize that Suzy's inspections will affect their own future in the company. This is because Suzy reports to management on the supervisors' compliance with internal safety standards. Unsatisfactory findings are considered to be adverse performance reports for the supervisors and their superiors. Positive reports are considered along with other performance variables in promotion and salary increases. This is not the only measure of safety that management depends on, but it is one factor in the determination of a given supervisor's or manager's future in the company.

One of the keys in the whole process is to take corrective action based on the findings. It sounds so obvious, yet audits and inspections are performed regularly with recommendations reported to management, but many are never acted on. Typically, following a major disaster that results in loss of lives and property, workers stand around saying they knew it was coming. "We tried to tell management, but no one listened." Often the problems were picked up in the regular audit or inspection procedure, recommendations were made, but no corrective actions took place. The safety practitioner must successfully sell management on the necessary course of action.

DIRECTING

The safety practitioner does not actively direct or lead in the organization, unless she has a staff or resources of her own. This job belongs primarily to line management. Ideally, the safety practitioner will be given adequate resources, including a safety budget, that can be used to help line management accomplish its objectives. A typical safety budget would be two to three percent of sales volume.

One day, Carl Wagner approaches Suzy Harris with a safety problem. He believes that it would be safer if his employees had hard hats to wear when they are working in "the yard." The cost of supplying the hard hats for the whole crew is $600. Suzy believes that this is a worthwhile project and agrees to pay for the hard hats. The fact that Suzy has control of a budget puts her in the enviable position of having managers come to her to request not only her verbal input, but also financial input. Safety personnel operating from a staff position will only be successful with support, including financial support, from line management.

STAFFING

Staffing is an issue that organizations should embrace. The opportunity to hire productive, creative people is also conducive to the growth and advancement of the whole organization. In addition, the safety practitioner should be aware that the process contains pitfalls. The Civil Rights Act and the Americans with Disabilities Act (ADA) have requirements that prohibit discrimination against protected populations. If a job requires strength and endurance, the safety practitioner should ensure that the company has a clear, written job description. Applicants should only be hired contingent on their ability to meet all pre-determined physical requirements. Review carefully all affirmative action regulations to assure that company screening procedures are in compliance with the acts. Many well-meaning companies have found themselves subject to lawsuits from state human rights commissions because they were unaware that their hiring practices discriminated against protected populations. Specific guidelines should be obtained from the appropriate state human rights commission.

COMMUNICATIONS

The ability to communicate effectively is critical to the success of safety practitioners. They must be able to speak in terms that management understands. This requires a knowledge of accounting, economics, and modern production and quality theory. Strong human-relations skills and related language ability are important to any successful safety effort. The safety practitioner will be working with top management and front-line workers. He needs to have the personality and ability to relate to both groups effectively.

The typical safety practitioner spends significant time in front of groups, often in training activities. Public speaking skills can be useful in these situations. Thoughts need to be well organized and the presentation should always be polished. This requires preparation and practice. Safety professionals who are not properly prepared in public speaking may join professional groups or they may take courses to help them prepare. In any case, they must learn to effectively communicate in a variety of situations and with a variety of groups.

EVALUATION OF THE SYSTEM

Since the safety practitioner is operating from a staff position, any success will be a result of his ability to enlist the support of line personnel. This comes about as a result of being well integrated into the organizational structure and culture, and it is also a result of being able to enlist the support and cooperation of the line managers. Obviously, two things are needed. First, top management must have already endorsed safety as being important to the organization, and it must have already given safety an appropriate level of support. This occurs because management perceives that safety is a worthwhile and contributing entity within the organization and that its activities are cost effective. Management will provide support for safety to the level and degree that it perceives that it is getting a reasonable return from its investment.

Support not only comes in terms of resources, but also in terms of commitment to holding all members of the management team responsible and accountable for safety within their own operations. Each and every manager is responsible, not only for production, but also for safe production. The manager's performance is measured relative to how well he performs in a manner leading to the safe production. Within the organization, the results of this measurement are used in evaluating the manager for promotion and salary.

Second, the safety manager must be perceived to be an integral part of the management team. Line managers can call upon the safety manager for advice and help in making their own operations safer. The safety manager's guidance will help them to create and maintain a safe workplace for their employees. Line managers can make proposals and compete for the safety staff's time and budget to help them, but, ultimately, the responsibility and accountability for safety rests on their shoulders. Implementing the safety program is difficult because the line manager must believe that the safety staff can help him accomplish his objectives. The safety manager and his staff become ex-officio members of his team. When he meets with his staff to discuss safety problems, he may request that a member of the safety staff participate in those sessions.

A serious mistake that some organizations make is believing that they can have two separate systems for evaluating production and safety. This occurs when no measurement for safety is built into the normal evaluation system and a safety incentive program is initiated to reward safe performance. The line manager is rewarded in terms of salary and position for having an exemplary production record. At the same time, he is punished by not receiving the perks associated with having no incidents or no lost-time accidents or whatever else the quota of the month is. The mixed signals do little to encourage safe production, but, instead, give the manager incentive to cover up any information that he perceives will inhibit him from receiving the reward or bonus associated with the safety program.

OSHA GUIDELINES

On January 29, 1989, OSHA published voluntary safety and health guidelines for general industry. OSHA "concluded that effective management of worker safety and health protection is a decisive factor in reducing the extent and the severity of work-related injuries and illnesses. Effective management addresses all work-related hazards including those potential hazards that could result from a change in worksite condition or practices. It addresses hazards whether or not they are regulated by government standards."

In general, OSHA advises employers to maintain a program which provides a systematic approach to recognize and protect employees from workplace hazards. This requires the following:

1. Management commitment and employee involvement are complementary. Management should value worker safety and health and commit to its visible pursuit as it would other organizational goals. A means should be established to encourage workers to develop and /or express their own commitment. This requires clearly stating and communicating safety and health policies and holding managers, supervisors, and employees accountable for meeting their safety responsibilities.

2. Worksite analysis involves examining the workplace for existing and potential hazards. Comprehensive baseline and periodic safety and health surveys should be conducted. Job hazard analysis, accident and near-miss investigations should also be held. Workers should be able

to report unsafe conditions without fear of reprisal. Trends of illness and injury should be studied over time to identify patterns and prevent problems from recurring.

3. Once hazards or potential hazards are recognized, they should be controlled, prevented, and/or eliminated. This requires engineering controls where appropriate, administrative controls, or personal protective equipment where necessary. Emergency plans, complete with drills and training, should be evaluated. Medical programs should be established.

4. Training should address the safety and health responsibilities of all personnel. Managers, supervisors, and workers should understand their responsibilities and the reasons behind them. This training should be reinforced through performance feedback and enforcement of safe work practices.

CONCLUSION

Because the safety practitioner is operating from a staff position, any success will be a result of his ability to enlist the support of line personnel. This comes about as a result of being well integrated into the organizational structure and culture and as a result of being able to enlist the support and cooperation of the line management. Two things are needed. First, top management must have already blessed safety as being important to the organization, and it must have already given safety an appropriate level of support. This blessing occurs because management perceives that safety is a worthwhile and contributing entity within the organization and that its activities are of cost benefit. Management will provide support for safety to the level and degree that it perceives that it getting a return from the same.

Support not only comes in terms of resources, but also in terms of commitment to holding all members of the management team responsible and accountable for safety within their own operations. Each and every manager is responsible, not just for production, but for safe production. The manager's performance is measured relative to how well he performs in a manner leading to safe production, and the results of this measurement are used in evaluating the manager for promotion and salary within the organization.

Second, the safety manager must be perceived to be an integral part of the management team. Line managers should be able to call on the safety manager for advice and help in making their own operations safer. The safety manager's guidance will help them to create and maintain a safe workplace for their employees. Line managers can make proposals for and compete for the safety staff's time and budget to help them. However, the responsibility and accountability for safety rests on their shoulders. Implementation difficulties arise from the fact that the line manager must have reason to believe that the safety staff can help him accomplish his own objectives. The safety manager and his staff become ex-officio members of his own team. When he meets his staff to discuss safety problems, he may request that a member of the safety staff participate in those sessions.

Questions

1. Do you think that safety practitioners should ever be given more than simple staff advisory authority? Why or why not? Do safety practitioners run the risk of assuming others' safety responsibilities when they do?

2. If there is a conflict between safety goals and production goals, which should take precedence and why? Ask the same question while playing roles of different individuals in the organization such as different types of managers. Also assume roles of interested parties outside the organization such as stockholders, bankers lending to the company, and government officials.

3. What steps should management take to assure employee involvement? Should any of these steps be legal requirements?

4. How do safety practices today compare to hose several decades ago? Talk to some "old timers" in industry and ask them if they are aware of any safety "cover-ups" that occurred in their industry. Discuss your responses.

5. What are the opinions of younger and older working adults about safety on the job, OSHA, and other safety-related issues. Do an informal survey. If a difference of opinion exists between the two groups, why do you suppose that is? As a manager, how would you approach each group to get its involvement?

Reference

Department of Labor. 1989. 29 CRF part 1910. *Federal Register*. 54. pp. 3909-3910.

Chapter 11

Psychology and Safety:
The Human Element in Loss Prevention

CHAPTER OBJECTIVES

After completing this chapter, you be able to

- Define the terminology associated with the study of psychology.
- Explain the concepts associated with motivation and safety.
- Differentiate between Behavioral and Goal-Directed theories of motivation.
- Explain the principles associated with behavior modification and safety.
- Describe the importance of establishing a positive safety culture.
- Identify the pitfalls inherent in safety incentive programs.
- Describe the benefits of employee empowerment and job enrichment.

CASE STUDIES

An electrical maintenance supervisor with fifteen years experience ordered his assistant to deenergize a machine that they were about to service. Before the assistant could "cut the power", the supervisor reached into the back of the unit, made contact with bare leads, and was electrocuted. The supervisor was an outstanding employee with no record of accidents or related lost work days. He had violated several company safety policies, however, while performing this tragic task. An electrician with this individual's years of experience should never have made these mistakes. The results of the accident investigation revealed that the supervisor was having marital problems. In addition, his spouse had indicated that just days before the incident the victim had shown her the location of his life insurance policies.

A small Midwestern manufacturing facility reported that 23 of their 75 employees had experienced back injuries. A job satisfaction survey indicated very low employee morale and workers that did not trust management. The survey results also indicated that management was only concerned with profits and production and not employee safety.

A company with a terrible safety record started a safety bingo program. Every week that a department did not report a recordable accident, members of that department could randomly select "bingo" stickers. Stickers were attached to a card that could be submitted for prizes when a column, row, or diagonal was completed. Safety performance was excellent until the awareness program ended. Injuries increased and production decreased following the termination of the incentive program.

INTRODUCTION

The safety professional in an organization has an extensive list of responsibilities. One of the most important of these responsibilities is the understanding of human motivation, capabilities, and limitations. Numerous safety researchers have pointed out that between 85 and 95 percent of all accidents are the result of human error. There is little doubt that people contribute to the hazards that result in injury or illness.

An important question must be asked given the fact that it is impossible to have a risk-free workplace. Why do only select individuals get hurt or become ill, although many individuals may be exposed to the same hazards? In addition, why do problems off-the-job affect performance on the job? Physiological and psychological differences among workers may be one explanation. It is useful for safety professionals to become familiar with the differences and similarities between people to help them develop effective strategies for loss control. When studying psychology and safety, most people think of personality (mental states) and how it affects the likelihood of injury. A review of the recent literature reveals that there are other areas that safety professionals and social science researchers are currently studying. Historic topics that are still of importance today include motivation and injury prevention, job satisfaction and safety, attitudes and safety, personality and loss prevention, and management leadership styles and injury prevention. Most recently, violence in the workplace, the safety culture of an organization, incentive programs, discipline and safety, group dynamics (teams) and injury prevention, and behavior-modification applications in health and safety have gained attention.

This chapter will superficially investigate several of these topics. Traditional approaches to the study of motivation will be examined to assure that the reader has a basic understanding of this topic. Various theories of motivation will be presented to assist the professional in recognizing some of the factors that contribute to worker behavior. At the same time, information presented in this chapter will expose the safety professional to various psychological concepts and theories. The intent is to provide useful strategies for correcting or eliminating psychological safety factors from both the worker and employer perspective.

This chapter will also focus upon the employee as a rational and mature adult. Quite often, simplistic solutions are applied to what is considered simplistic behavioral problems. Recent literature has pointed out that many of these solutions may be eliminating some behaviors while actually creating more significant and difficult underlying problems. A "holistic" approach to the human element in occupational safety and health will be examined by discussing the safety culture of an organization. In addition, incentive program pitfalls will be discussed along with the impact of employee empowerment and job enrichment on safe task performance.

BASIC TERMINOLOGY

Before delving into the topic of psychology and safety, several key terms should be defined. The term "psychology" can be defined as the study of behavior. It is a wide area of study including clinical psychology, developmental psychology, educational psychology, experimental psychology, industrial psychology, social psychology and physiological psychology.

When examining the area of human behavior, several psychological terms are often mentioned. Some psychologists believe that mental states influence behavior and that attitudes are associated with mental states.

Attitudes are enduring reactions toward people, places, or objects, based on our beliefs and emotional feelings. For example, what if we are raised to believe that tall people are complainers? When a tall person hurts his back, what is our reaction likely to be? "That tall person is exaggerating his injury; he is not really hurt. There was no problem with the job; he should have been able to lift the 180-pound load. If he really wants to work, he has to accept all of the responsibilities of the job." Notice how this attitude influenced the way that the injury scenario is viewed. The question of bias is especially important when safety professionals are concerned with hazard identification, incident trend analysis, and accident investigation. Individuals' attitudes may guide their attention and efforts in the wrong direction.

In the occupational environment, *job satisfaction* is another mental state that is thought to influence workplace behavior. Job satisfaction is a specific attitude. It is the emotional feeling that individuals have about their jobs. When workers enjoy what they do, they are thought to be receiving satisfaction from the job. When work experiences are negative, such as conflicts with bosses or coworkers, workers may be receiving very little job satisfaction. Negative job satisfaction may then influence employee morale.

Morale is considered to be the satisfaction of individuals' needs and the extent to which employees recognize that this satisfaction comes from their jobs. Some researchers believe that if morale is low, employee motivation will also be low. *Motivation* may be viewed as an inner drive, impulse, or need that creates a personal incentive toward behavior. Some psychologists explain workplace behavior in terms of motivation.

Other psychologists, however, only look at worker behavior and the factors that influence the behavior. Behavior, which includes both visible action, as well as physiological responses, is the primary focal point for these scientists. These behavioral researchers are more concerned about the environmental factors that influence performance then the mental states which are considered difficult to accurately measure. *Behavior modification*, the application of operant conditioning principles for behavioral change in the occupational environment, is an area becoming increasingly popular in the safety profession. Behaviorally oriented topics will be discussed in greater detail later in this chapter.

MOTIVATION

Will people do something just because they know it is right? Obeying speed limits on our nation's highways, wearing seat belts while operating our cars, and using personal protective equipment while performing hazardous jobs are sensible rules that many individuals frequently decide not to follow. Why do people violate rules when they know that their actions are not right?

Similar remarks are frequently made by safety professionals in the occupational safety environment: "How can I get employees to work safely? What do I have to do to motivate them?" "My worker's have bad safety attitudes." When professionals make these comments, they are missing the point regarding the understanding of human behavior and performance. While there are numerous theories that attempt to explain why people behave the way they do in given situations, there is one universal and underlying point of agreement. By definition, *motivation* can be described as an individual's tendency toward action in a given situation. In other words, people cannot and do not motivate others. It is the individual who acts or behaves in a given situation. Environmental conditions can be established to increase the likelihood of action and performance, but it remains for the individual to respond. How that individual actually responds depends on that individual's vast history of personal experiences. This is often described in terms of worker attitude. Depending on the particular school of thought, however, internal or external factors must be taken into account in predicting behavior. Two of the most popular views of motivation are represented by the *Goal-Directed* and the *Behavioral* schools of study. In the Goal-Directed School of Motivation, the inner drives of individuals are examined to explain why human behavior takes place. Examples of goal-directed theories include The Needs-Hierarchy Theory, The Need Achievement Theory, and The Motivational Hygiene Theory. Examples of Behavioral theories are Pavlovian theory and Operant theory.

The goal-directed study of motivation states that people have various types of needs. Depending on the specific theory of interest, these needs can be physiological (the need for food, clothing, or shelter) or psychological (the need for love, recognition, and affiliation with others). Needs help individuals to establish goals which guide their behavior to gain rewards and to satisfy those needs (refer to Figure 11-1: the model first developed by Deci that pictorially describes the relationship between needs and behavior).

NEEDS ⇒ ESTABLISHED GOALS ⇒BEHAVIOR ⇒ REWARDS

Figure 11-1: Goal-Directed model of motivation based upon the research (Deci.)

Dr. Abraham Maslow's *Needs-Hierarchy Theory* is based on the goal-directed model of motivation (refer to Figure 11-2 for a pictorial representation of Maslow's model of motivation). According to Maslow's theory of motivation, everyone has basic needs that must be satisfied before more advanced needs can become a motivating influence. In this theory, the basic physiological needs required for fundamental human survival must first be obtained. Hunger, thirst, and warmth are essential needs that motivate people to seek food, water, shelter, or clothing. It is only after these basic physiological needs are satisfied that individuals seek safety and security. Safety and security refer to those needs that protect and preserve the health and well-being of the individual. Living in a location free from turmoil and conflict is an example of this level of need.

Once the safety and security needs are satisfied, individuals seek to establish close friendships and loving relationships with others. With this social support system in place, the individual then strives to satisfy those needs associated with self-esteem. Self-esteem refers to the individual's sense of personal worth. Self respect, dignity, independence, and confidence are some of the needs that individual's are trying to achieve at this level.

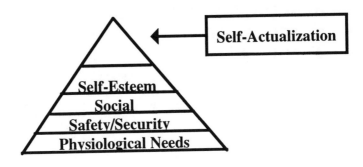

Figure 11-2: Maslow's Needs-Hierarchy Model of motivation.

Maslow theorized that the individual, having satisfied all the lower levels of needs, now had the potential for self-actualization. Self-actualization refers to an individual who is more concerned about the well-being of others or the state of society. Maslow's examples of self-actualized individuals, exceptional individuals in any society, include Eleanor Roosevelt and Mother Teresa.

There is little evidence to justify the accuracy of this theory in explaining human motivation. Maslow's work is based on post-hoc studies of selected individuals. It is important to point out, however, that researchers have found that certain needs must be satisfied before individuals can or will pursue other needs. Any of Maslow's needs could serve as motivators for individuals depending on their psychological state. A goal-directed theory that has acquired significant attention because of the tangible research that support its premise is McClelland's *Need Achievement Theory*.

According to McClelland, there are three essential motives that drive human behavior:

- **ACHIEVEMENT**
- **AFFILIATION**
- **POWER**

Achievement is described as the accomplishment of tasks and activities. The sense of accomplishment and the satisfaction of the need to achieve, according to this theory, is obtained from the completion of a task.

According to the research in this area, individuals who have a high need for achievement are very comfortable working alone. They tend to be very creative individuals who can develop unique solutions to the challenges presented to them.

The second need, the need for *affiliation*, points to the importance of individuals seeking out the company of others. Individuals with a high need for affiliation often gain satisfaction from knowing or meeting famous people. Often, however, this need is satisfied by close relationships with others. The enjoyment of being a part of a certain work crew and talking with co-workers would be methods which are used to satisfy this need. Pride, associated with the wearing of team or company clothing, is an example of the outward display of affiliation.

The need for *power* is probably the most misinterpreted of McClelland's three needs. Power, in this context, does not necessarily mean insidious and tyrannical control over people. McClelland views power as the ability to positively influence others. Managers often have high power and achievement profiles. They have a need to get tasks accomplished through the use and influence of subordinates in the organization. According to this theory, many teachers and trainers have a high need for power. They gain great satisfaction from student successes and accomplishments.

While some individuals may have a predisposition toward one of the three needs, it is more likely that they will possess several of these needs. McClelland points out that it is useful to be aware of individual differences when trying to understand motivation. Frequently, we attempt to influence the behavior of others by using rewards that are important to us. Remember, it is only possible to influence the behavior of others when you know what is important to them.

The last goal-directed theory that will be presented is the *Motivation Hygiene Theory*. In the motivation hygiene theory, the employee motivation depends on the characteristics of the job. These job characteristics are referred to as intrinsic job factors. There are fourteen factors: achievement, activity, authority, creativity, importance, independence, interest, knowledge, personal growth and development, promotion opportunity, recognition, responsibility, service to others, utilization, and variety. Let's examine a few of the less-obvious intrinsic job factors.

Activity refers to a steady and physically acceptable level of performance. The important issue here is that physical and mental workloads are balanced and reasonable. When workloads are either excessive or minimal, occupational stress resulting from fatigue or boredom can be a safety problem.

Authority refers to the amount of power that is inherent in a job to influence or control the work activities of others. Here again, occupational stress is often attributed to positions where individuals have extensive responsibilities but little control over the accomplishment of tasks. Job satisfaction is closely tied to the appropriate balance of responsibility (accountability for performance, decisions, and outcomes) and authority. *Creativity* is the intrinsic factor reflecting the opportunity for individuals to apply inventiveness, resourcefulness, and personal talents to the work situation. Many individuals enjoy the process of developing new and innovative solutions to problems in the workplace. Jobs that tap into an individual's creativity are often seen as more enjoyable because they tend not to be routine.

Importance refers to the sense that the work being performed is valuable. When employees feel that they make a difference in the organization, job satisfaction increases for those workers. When coupled with the intrinsic factor of *interest*, the chance to perform activities that are compatible with the individual's personal preferences or interest, employee motivation tends to soar. Adding promotion opportunity; the likelihood of advancement and recognition; and company appreciation of individual performance to these factors, makes motivation very high.

Remember, intrinsic job factors will only be motivational if the individual considers them as such. Not everyone desires advancement or promotion. As a consequence, this intrinsic job factor may not be motivating to the individual who desires to remain a member of a crew or team.

It must be pointed out that the concept of job enrichment is based on the motivational hygiene theory. The greater the intrinsic job factors that a position possesses, the greater the job satisfaction. *Job satisfaction* is a term used to describe the positive feeling that workers have about their jobs. It implies that workers are satisfying many of their personal needs through their work. Several studies indicate that a negative correlation exists between job satisfaction and accidents. Research has reported that the greater the job satisfaction of employees in a company, the fewer the reported accidents.

The behavioral school of motivation is quite different from the goal-oriented school of motivation theory. The behavioral school examines the environmental factors that influence human performance. Two of the most widely known behavioral theories are the Pavlovian theory, also know as the *classical conditioning* theory, and *operant conditioning* or Skinnerian theory. The difference between these two theories is that classical conditioning views behavior as reflexive while the operant conditioning views behavior as learned and dependent on environmental consequences. Pavlovian conditioning explains behavior as being reflexive in nature. The traditional example is of dogs salivating when a stimulus, a bell, is paired with the presentation of food. After repeated pairings, the "conditioned" or learned stimulus produces the same response as the "unconditioned" or natural stimulus. An example of this type of motivation is the fear and anxiety people feel when they enter a location where they have been injured. The location becomes a conditioned stimulus or cue producing the natural emotional response.

People involved in serious automobile accidents often experience anxiety when they attempt to start driving again. One industrial example is the near fatal electrocution of an electrician at a high-voltage substation. When the electrician returned to work after a six-month absence, the employee could not pass through the gates of the power substation without experiencing debilitating anxiety. The power substation became his conditioned stimulus which was associated with severe and intense pain.

Operant conditioning, on the other hand, is a process in which the frequency of occurrence of behavior is modified by the consequences of the behavior. The fundamental concept is that the consequences of people's actions will determine future actions. These external consequences, referred to as the environment or environmental stimuli, may increase, decrease, or have little effect upon behavior. When the likelihood of a response increases following the presentation of an event or stimulus, the process is referred to as *positive reinforcement*. Praise, recognition, and financial rewards are just a few possible examples of positive reinforcers. By definition, however, a stimulus like money is not a positive reinforcer unless it increases the likelihood of the behavior reoccurring in the future.

When the likelihood of a response decreases following the presentation of an event or stimulus, then the process is referred to as *punishment*. Figure 11-3 is a chart that shows the relationship between the presentation and removal of environmental events and their effect upon behavior. Supervisor praise for employee compliance with a company safety hard-hat policy is an environmental stimulus that can only be considered a positive reinforcer if it increases the likelihood of the employee wearing the hard-hat in the future.

	STIMULUS	
BEHAVIOR	PRESENTATION	REMOVAL
⇑	**POSITIVE REINFORCEMENT**	**NEGATIVE REINFORCEMENT**
⇓	**PUNISHMENT**	**EXTINCTION**

Figure 11-3. Operant Conditioning model of environmental processes and their effect on behavior.

Many individuals confuse the process of punishment with negative reinforcement believing that both processes are one and the same. According to strict definitions used in operant conditioning, this is not the case. *Punishment* is the decreased likelihood of a behavior reoccurring as a result of the presentation of a stimulus. If employees are suspended every time they are observed performing unsafe behaviors and, over time, the unsafe behaviors decrease in frequency, this process would be considered punishment. *Negative reinforcement*, however, is the increased likelihood of a behavior to remove or avoid the presentation of an aversive stimulus. If the employee only wears the safety gear when the supervisor is in the work area, this behavior would be considered *avoidance*, being maintained by negative reinforcement. This is an example of negative reinforcement because the employee is avoiding the possible negative consequences of punishment.

Many people also confuse the processes of extinction and punishment. *Extinction* is the removal of a positive stimulus. This reduces the likelihood of behavior reoccurring in the future. An example of the extinction process would be the elimination of incentive programs. Incentive programs are often used to increase safety performance. Frequently, when incentive programs are eliminated, safety performance will decrease. The reduction in desired safety performance would be explained by behaviorists as resulting from the process of extinction. This is a lesson that safety professionals must keep in mind. Before starting incentive programs such as safety awards, prizes for safety competitions, or other positive incentives, several questions must be answered. How long is a company prepared to maintain incentives? How much is the company willing to spend? What happens when budget cuts must be made? Once rewards are awarded, employees view them as entitlement. Remove these entitlement and the company will face significant anger and disappointment from its workers. In addition, negative side effects may result from safety competitions. Issues and concerns regarding incentive programs will be discussed at the end of this chapter.

THE RATIONAL EMPLOYEE: APPLYING MOTIVATION THEORIES

Several motivation theories exist that attempt to explain employee behavior. There is one underlying message that the safety professional should recognize based on this information. Employees are rational. They have specific reasons for behaving the way they do. The challenge will be to determine those factors that are supporting undesirable performance and develop strategies to modify that behavior. In today's workplace environment, employee participation and team building may be the motivational strategy to implement.

As you have probably noticed, based on the succinct review of the theories presented in this chapter, there are several possible explanations for undesirable safety performance. Factors that must be considered include employee background, peer influence, and company environment. One or more of these factors may be supporting the undesirable behavior and must be eliminated.

Suppose that a new employee must work near a vat of acid. The employee has been trained in recognizing hazards in the workplace but still does not wear personal protective equipment (PPE). Is it irrational not to wear eye protection near acid? Most people would probably say, "Yes!" This undesirable behavior seems to be irrational, but is it as irrational as it first appears? If coworkers heckle the new employee for talking about safety and wearing PPE, which factor becomes more important? From the need achievement theory perspective, the new employee may have a need for affiliation. If the need for affiliation with coworkers is high, the new employee will choose to follow the norms established. This scenario can also be examined from the perspective of the Operant Conditioning theory. Coworker verbal abuse might be considered as a punishing stimulus. If the new employee is punished every time he is observed wearing eye protection and his supervisor ignores the safety rule violation, what is the likely result?

There is another explanation for the persistence of unsafe behavior. What is the organization's response when safety rules are violated or unsafe practices occur? The safety professional must look at the organizational environment that is maintaining this performance. While safety rules and polices may be written, there could be inconsistent enforcement of the rules. In the same safety scenario, the new employee will compare the importance of the social reinforcer of "hanging out with the crew" with the likelihood of punishment from the company. The selection of the social reinforcer is especially likely if the discipline is only a slap on the wrist. An even worse situation would be the inconsistent enforcement of company safety rules. If workers are not reprimanded when they are observed violating safety policy, how is the employee likely to react? Resulting employee behavior is far more rational then we might initially believe.

ORGANIZATIONAL ENVIRONMENT AND THE SAFETY CULTURE

It is important to mention that, from a behavioral perspective, organizational actions are just as important as organizational directives. In a Fortune 500 company, for example, senior management decided that it was good employee relations to emphasize safety. Safety slogans were visibly posted throughout company facilities. Weekly "tool box" safety meetings were mandated for company work crews. Safety messages were included in all employee payroll envelopes. The company went "all out" to promote safety awareness.

During programmed maintenance shutdowns at this same company, however, employees were instructed to get equipment running quickly. The underlying message to the employees was get production moving no matter what the cost. Employees were informed that approximately one million dollars were lost every day that this particular plant was not operating. They understood the message. Several mechanics and utility workers were seriously injured during the maintenance shutdown. Employees took short cuts and failed to use the correct tools and procedures to perform many of the tasks. Wearing of PPE was also not enforced.

This example is not unique. An organization's actions, along with the types of statistics that it measures, clearly communicates what is important. Several authors have pointed out that when safety performance is measured in terms of injury statistics, there is a tendency for employees not to report incidents. A company may look good on paper when fat/cats (fatalities and catastrophic accidents) are lurking in the corner.

Another factor which must be considered is the motive behind an organization's actions. Company motives are more transparent to employees than some managers believe. Responding to the fears associated with OSHA inspections and citations will not build a safety culture that will be supported by employees. If employers are pressured to reduce insurance costs by lowering accident frequency and severity rates, employees will recognize when budgets are provided for quick fixes and when they are provided for sustained processes.

The term *safety culture* has been mentioned in several recent articles, but what does it really mean? According to *Webster's Universal Dictionary*, the word culture has Latin roots associated with agriculture, cultivation, nurturing, and care. It has several possible definitions, but the most appropriate one is, "Education, training; balance of intellect and judgment." When considered from a safety management perspective, a safety culture should be viewed as the nurturing and cultivation of an organizational value system that guides all organizational actions. It is the balance of business intellect and employee moral judgment.

In his bestseller, *The 7 Habits of Highly Effective People*, Stephen Covey discusses the relationship among paradigm, values, principles, and success. As Covey points out, a paradigm is a model or frame of reference. It is used to understand "what is happening at any given point in time," as well as "where we want to go." Organizations need a culture or paradigm that will help them interpret events that occur at any given point in time. They also need a culture with clearly specified goals of what the organization wishes to achieve. An organizational paradigm along with a clear conceptualization of values, principles, and goals is the foundation of a culture that will promote the right actions to achieve excellence and success.

A value system is a set of convictions that a person holds about a specific mode of conduct and the relative importance of those convictions. Value systems and principles serve the same purpose for organizations. In other words, success depends on what is important to the organization. What is thought to be important is based on what others have taught and what is learned. What an organization considers important is dependent upon what corporate management considers important. It is ultimately based on the lessons learned by key individuals in a company. For both organizations and individuals, these are values.

An example of the influence of organizational paradigms, values, and safety cultures on management decision making was observed in a Fortune 500 company's construction project. A textile manufacturer was constructing a new facility. Upper- level management, including the safety and operations manager, agreed that all contractors would have to comply with all company health and safety rules during the construction project. In all of the written contracts, this was clearly noted. If a violation was observed by the textile manufacturer, the employee of the subcontractor would be removed from the job site. A repeat violation meant the termination of the subcontractor's contract.

Six months into this two-year project, one of the major subcontractors responsible for multi-story steel-beam erection had employees violating safety policies. This subcontractor had several employees working over 30 feet in the air without fall protection. It was the second observation of this violation. The plant safety manager ordered the workers down, informed the contractor of the repeat violation, and initiated the process to terminate the contract.

What was the response of the plant manager to this incident? In companies without clear and firmly established paradigms, a plant manager might compromise safety values. The plant manager might stretch the rules to get the project back on schedule. The plant manager in this scenario did not compromise. The contractor was removed from the project and replaced by a firm that accomplished all the goals initially established by the textile manufacturer. This project was completed on time with only one OSHA-recordable injury.

Values influence behavior. They impact the way people work, the way people treat others, or the way people react to figures of authority. Substitute the word organization for people and it can be seen that this paradigm also holds true for organizational cultures. Values are not part of the paradigm or guidelines that magically appear over night. For most people their parents, brothers and sisters, sports heroes, teachers, friends, and religious leaders are just a few of the many individuals who influence their values. For organizations, it is employees' values that mold the organizational values and establish, whether consciously or unconsciously, the safety culture and how it defines success.

It is critical for the safety professional to work with the organization and its management to establish the safety paradigm and safety culture. The safety professional must cultivate the knowledge of those in influential positions, taking advantage of every opportunity to educate management on the benefits of a safety culture. Gathering necessary data, knowing the regulations, and presenting examples of what other organizations are doing are some of the topics that should be covered. The other area that must be cultivated includes the values and principles of the organization. Sometimes actions must be taken because they are the right things to do. This will only be possible if the organization has clearly established values and principles in place. The safety professional's goal must be to assist in the development of those values and principles.

BEHAVIOR MODIFICATION AND SAFETY

Articles discussing behavior modification and safety have appeared in ever-increasing numbers since Komaki and Sulzer-Azaroff first published their ground-breaking research in 1978. However, while the term behavior modification is frequently used, it is often used inappropriately. *Behavior modification* refers to the application of the principles of operant conditioning theories of learning to obtain relatively permanent changes in the way that individuals respond to specific situations. As has already been discussed in the section on motivation, these principles involve the environmental influences on behavior.

Behavior modification examines three key elements of behavior. These three elements are referred to as the ABCs of behavioral change:

A =   Antecedents
B =   Behavior
C =   Consequences

Antecedents refer to the stimuli or cues in the workplace that prompt a behavior. They can be persons, places, or objects that indicate that specific types of behavior should be emitted. *Behavior*, in the operant model, is defined as anything that is observable and measurable. Consequences are the events in the environment that influence the behavior. An example of the ABC model would be an employee entering a noisy area that has been posted with a yellow "wear hearing protection" sign. The noise and the sign serve as antecedents for the desirable worker behavior of wearing ear plugs. In this example, the consequence could be a reduction of pain because of the decrease in sound intensity. Another consequence could be praise by a manager observing the desirable behavior.

Behavior modification programs that have been successful in increasing desirable behavior and reducing injuries have several common characteristics:

1. Tasks are divided into small behavioral units and the desired behaviors are specified.

2. The desired behaviors are clearly communicated and adequately taught.
3. Based on objective measurements, there is regular and continued feedback of the performance level being achieved (Saari, 1994, p. 11).

This three-stage process is typically started after a task analysis has been performed. This process can be referred to as using behavioral menus. It is possible to use a Job Safety Analysis (JSA) to provide the task's steps (refer to the chapter on Systems Safety for an explanation of how to conduct JSAs). Training programs are then conducted with clearly specified criteria for performance evaluation. The feedback mechanism is to show performance on a chart in terms of frequency of behavior occurrence. It should be pointed out that safe performance, a positive measurement, is the behavior of interest.

In most of the successful behavior modification and safety articles published to date, the consequence of desired behavior has been the feedback of performance. Some authors have recently suggested the use of incentives to serve as positive reinforcement consequences. This author believes that there is real danger in using these incentive programs.

INCENTIVES VERSUS INHERENT REINFORCERS

Safety award programs have been around for a long time. The premise behind the use of incentives is that employees require added encouragement to work safely. Many incentive programs use statistical measures such as recordable accidents on the OSHA 200 log or days without a lost-time accident to determine the winners. Awards have taken a variety of forms. They can be anything from ball caps, tee-shirts, or jackets to automobiles and trucks.

Several recent articles have recommended the use of incentive programs as a method to advance the safety culture of an organization. The rationale for this opinion is that management is demonstrating its willingness to go beyond a pay check to elevate safety. One article suggested that safety awards should be linked to the important activities of a total safety culture. By promoting activities rather than the traditional use of accident statistics, the company could get more long-term value out of the incentive program. While safety-related activities are more important than statistics that are beyond an individual's control, incentive programs should not be the method to demonstrate the importance of the safety culture.

It is an unfortunate fact, however, that incentive programs create more problems than they solve. Research repeatedly demonstrates that incentive programs improve safety while they are in place. Once the incentive program ends, performance will return to previous levels at best. In some instances safety performance is much worse. Employees often view incentive programs as entitlement. After a period of time, these bonuses are expected by the employees. Remember the discussion of extinction according to the operant conditioning theory of motivation. Remove a reinforcer and the behavior will terminate.

Incentive programs can also create unhealthy competition. Individuals, crews, departments, or plants will often undermine the activities of their competitors to gain the awards. At one company, the department with the cleanest work area during the previous month was awarded with a steak dinner. Reports of trash dumping in competing departments is just one example of how competition for incentives can backfire.

Incentive programs based on statistics often produce the unwanted side effect of inaccurate recordkeeping. At another company, employees had worked over eleven months without a lost-time accident. This was quite an achievement. It was thought to have been accomplished because of a new safety award program that was initiated early that year. Word circulated around the facility that the company CEO planned to fly to the facility and present the award for going one year without a recordable accident. Approximately two weeks before the big day an employee slipped while using a prybar. The prybar struck the employee in the face, shattering his safety glasses. While the employee avoided a serious eye injury he did receive a facial laceration requiring stitches. Under normal circumstances he would have been sent home after a visit to the emergency room. Instead, both supervisors and coworkers pressured the employee to return to work.

The facility had its award ceremony, but at what price? Yes, the incentive program motivated the employee to return to work. However, this a recordable accident according to OSHA record keeping guidelines. What does this really say about the company's safety culture?

Another pitfall of incentive programs is what is considered as a positive reinforcer by management, may not be to the participants of the incentive program. Coal miners at one facility had the opportunity to be in a drawing for a new pickup truck if they went one year without a recordable injury. Did this have an impact on a dismal safety record? No! When safety professionals checked into the failure of this incentive program, it was determined that employees were financially better off when they were off of work with an injury. When asked why the truck award incentive program failed, interviewed miners reported that they received workers' compensation benefits which matched approximately 70 percent of their normal salary. In addition, their homes, cars, boats, and other loans were automatically paid for by the loan insurance that went into effect when they could not work as a result of an injury. Remember that by definition, a reinforcer is anything that increases the likelihood of behavior. Winning a pickup truck, according to this definition, was not a positive reinforcer.

Often incentive programs are used in place of effective safety programs and positive safety cultures. Safety is not magic. An effective safety program cannot be developed overnight. The short-term success of incentive programs will turn sour just as quickly when the program ends.

EMPLOYEE EMPOWERMENT AND JOB ENRICHMENT

With the new emphasis on safety culture and Total Quality Management (TQM), progressive companies are recognizing the value of employee-driven safety programs. Terms like employee empowerment and job enrichment reflect the importance of intrinsic job factors for the promotion of safety.

Today's employees are not receptive to authoritarian styles of management. As discussed in the motivational hygiene section of this chapter, intrinsic job factors that are inherently reinforcing will motivate employees. Some authors view all external rewards as punishers. This is understandable based on the discussion of incentive programs. Incentive programs are extrinsic reinforcers. Extrinsic reinforcers are "artificially" incorporated into the organization to increase the likelihood of desired performance. However, remove the artificial reinforcer and natural behavioral activities return to the work environment.

Intrinsic reinforcements, however, are those positive aspects of our behavior and environment that are self-perpetuating. Why do people watch television? Viewing television is entertaining, relaxing, and enjoyable. It has intrinsic reinforcement. Intrinsic reinforcers can exist in all activities. In an organization, job enrichment and empowerment refer to the incorporation of intrinsic job factors into the work. Some individuals desire jobs with responsibility and authority. Other employees want their jobs to be interesting and important. The more motivational hygiene characteristics associated with a task, the greater the likelihood that employees will find their jobs satisfying and rewarding.

Through employee participation in the safety culture movement within an organization, companies are driving out the fear and breaking down the barriers that Deming discusses in his TQM theory. When employees feel secure about their contribution in an organization, they are more likely to suggest better ways of performing their jobs. When barriers are down, the competition that undermines the success of the organization is eliminated. By sharing responsibility, everyone becomes a player with a vested interest for the success of the process. An understanding of human and organizational behavior can only increase your success as a safety professional.

CONCLUSION

This chapter superficially reviewed the theories of motivation. It examined the importance of the organizational environment and the safety culture. It examined incentive programs and their strengths and weaknesses. Finally, job enrichment and employee empowerment were discussed as methods to increase employee buy-in to the safety process. References at the end of this chapter can assist the safety professional in learning more about psychology and safety.

Questions

1. How does job satisfaction influence an individual's safety? What role does worker attitude play in accident prevention?

2. Why is knowledge of worker motivation important to a safety professional?

3. How are goal directed theories different from the behavioral theories of motivation?

4. How can the motivational hygiene theory of motivation be used to improve a safety program?

5. What is meant by the term safety culture? How do you establish a positive safety culture?

6. How can positive reinforcement and punishment be used in a behavior-modification-oriented safety program? Give examples.

References

Catania, A. C. 1968. *Contemporary research in operant behavior.* Glenview, IL: Scott, Foresman and Company.

Covey, R.S. 1889. *The 7 habits of highly effective people.* New York, NY: Simon & Schuster.

Daniels, A. C. 1989. *Performance Management: Improving Quality Productivity Through Positive Reinforcement.* Tucker, Georgia: Performance Management Publications.

Everly, G. S. and Feldman, R.H.L. 1985. Occupational health promotion. *Health Behavior in the Workplace.* New York, NY: John Wiley & Sons.

Geller, S. 1994, September. Ten principles for achieving a total safety culture. *Professional Safety, 39,* 19-24.

Henderson, C. J., and Cernohous, C. 1994, January. Ergonomics: A business approach. *Professional Safety, 39,* 27-31.

Kamp, J. 1994, May. Worker Psychology: Safety management's next frontier. *Professional Safety, 39,* 32-38.

Katz, D., and Kahn, R. L. 1978. *The Social Psychology of Organizations.* New York, NY: John Wiley & Sons.

Kohn, A. 1993. *Punished By Rewards.* Boston, MA: Houghton Mifflin Company.

Krause, T.R. 1995, February. Driving continuous improvement in safety. *Occupational Hazards, 57,* 47-49.

Krause, T.R 1995. Employee-driven systems for safe behavior: *Integrating Behavioral and Statistical Methodologies.* New York, NY: Van Nostrand Reinhold

Krause, T. R., Hidley, J. H., & Hodson, S.J. 1990. *The behavior-based safety process.* New York, NY: Van Nostrand Reinhold.

Lenckus, D. 1994, October. Safety awareness alone not enough to sow the seeds for fewer injuries. *Business Insurance, 42,* 3, 15-17.

Liebert, R. M., and Neale, J. M. 1977. *Psychology.* New York, NY: John Wiley & Sons.

Makin, P. J., and Sutherland, V. J. 1994, May. Reducing accidents using a behavioural approach. *Leadership & Organization Development Journal.* 15, 5-10.

Minter, S. G. 1994, January. A Safe Approach to Incentives. *Occupational Hazards, 57,* 171-172.

Nirenberg, J. S. 1984. *How to Sell Your Ideas*. New York, NY: McGraw Hill.

Preston, R., and Topf, M. 1994, March. Safety discipline: A constructive approach. *Occupational Hazards, 56,* 51-54.

Reynolds, G.S. 1968. *A Primer of Operant Conditioning*. Glenview, IL: Scott, Foresman and Company.

Saari, J. 1994, May. When does behavior modification prevent accidents? *Leadership and Organization Development Journal,* 15:11-15.

Stein, D. G., and Rosen, J. J. 1974. *Motivation and Emotion*. New York, NY: Macmillan.

Weinstock, M. P. 1994, March. Rewarding safety. *Occupational Hazards, 56,* 73-76.

Chapter 12

Workplace Violence

CHAPTER OBJECTIVES

After completing this chapter, you will be able to

- Describe the extent of the workplace violence problem in the United States.
- Compare the frequency of workplace violence fatalities to other causes of death on the job.
- Identify the high risk work environments experiencing workplace violence.
- List several occupations where workplace violence statistics indicate a problem of concern for the safety professional.
- List some of the factors that contribute to workplace violence.
- Explain the importance of establishing a workplace violence prevention program.
- Describe some of the strategies that companies can use to prevent or minimize the effects of workplace violence.

CASE STUDY

Jane Doe walked to work on a January evening in 1992. She did not have a car because she could not afford one. Minimal skills and education left her vulnerable in a number of ways. Jane worked as a clerk in a combination gas station convenience store. It was approximately 1:30 a.m., when an unknown assailant walked into the store ostensibly to make a purchase, but in reality to rob the store. Maybe Jane knew the assailant, or possibly he was afraid that she would recognize him later. In either case, before he left, Jane was dead, the victim of multiple stab wounds to the chest and neck. A few hundred dollars was missing from the cash drawer. Jane's body was discovered when the next customer came in for a purchase; her assailant has yet to be found.

INTRODUCTION

Robbery and criminal acts are the primary motives for homicide at work, accounting for seventy-five percent of these deaths. Approximately half of all homicide victims were working in retail establishments such as grocery stores, restaurants, bars, and small gas stations. These traditional night-time retailers are the most vulnerable (Toscano and Windau, 1994).

Managers and small business owners who have done little or nothing in the past to protect their property and, more importantly, their employees from robbery will find themselves being attacked on all fronts. Not only are they more vulnerable to losses from criminal activity, but they will also find themselves increasingly subject to litigation and OSHA penalties. Charles Jeffries, OSHA Director from North Carolina, stated that many previously uninspected businesses are going to be surprised when OSHA shows up. He referred to fast food and other smaller retailers as potential targets (Charles Jeffries, personal communication, October 5, 1994).

The reasons are obvious and are based on the dangers faced by employees of night-time retail operations. Employees like Jane can and should be protected. Here's how it can be done.

As in any other area of safety, it is advisable to have a written plan outlining policies designed to deal with the anticipated problem. The plan should include procedures for cash management, handling customers, and generally minimizing the likelihood of robbery. More importantly, employees should be carefully briefed and drilled on what to do if the worst happens.

Planning should include acquiring and installing certain basic equipment. Many night-time retailers, particularly convenience stores and gas stations, can benefit from the following.

Installing a security camera capable of recording any robbery that might take place should be a high priority. This will not only act as a deterrent, but can also be a means to help prevent a particular robber from returning to your operation or any other operation if apprehended. As a complement to the camera, have height markers placed next to the entrance which display measurements from the floor. The camera and the clerk can use these to help identify assailants.

Incorporate a drop safe or other cash management device to limit employee access to large amounts of cash. Of course, conspicuous notices must be placed stating that the cash register contains small amounts of money, such as fifty dollars or less. Employees should be carefully instructed in procedures and make a conscious effort to keep cash on hand low (Florida Department of Legal Affairs, Division of Victim Services, 1993).

When feasible, a silent alarm should be installed which will notify police or a private security force that a crime is in progress (Florida Department of Legal Affairs, Division of Victim Services, 1993). Prompt response by the appropriate authorities may save a life.

Ensure the store and parking lot are well lit during all operating hours. Limit the number of entrances that open from the outside and avoid permitting employees to exit into poorly lit, unobserved areas. They might be assaulted on the way out and the assailant may use the door as a means of entrance.

Convenience stores and other night-time retailers can help protect employees by not permitting employees to work alone at night. Two clerks in a convenience store can help protect each other. Many times clerks are the victim of two would-be shoppers. One will distract the clerk while the other pulls a weapon, attacks, and disables the victim. If a second employee is not an economically feasible alternative, then protective enclosures for the lone attendant are a must.

In Florida, where state statutes require one of the two alternatives, store owners are turning to companies such as Pro Tec in Greenville, North Carolina, to help protect employees (Florida Department of Legal Affairs, Division of Victim Services, 1993, p. 13). All monies and the worker stay behind an open counter during the day and early evening. After traffic thins out and the store becomes more vulnerable, a laminated plexiglass shield made of Lexan is dropped into place from overhead. With plate steel behind and a clear shield between the employee and would-be assailant above the counter, a high level of protection is afforded against robbery and violence. In some cases, the unit is placed at the door of the store with a transparent turnstile permitting the employee to pass goods to and receive money from customers. If the owner believes this will inhibit the sale of merchandise from the store, then the enclosure can be placed around the employee inside the store.

These shields convert the night-time retail establishment into a veritable fortress in short order. The bullet resistant shield stops the penetration of clubs, knives, and typical small-caliber firearms. Under Florida law the enclosure must meet Underwriters Laboratory Standard UL 752 for medium-power small arms, bullet-resisting equipment, or the American Society for Testing and materials Standard D3935 (Classification PC110 B 3 0800700) (Florida Department of Legal Affairs, Division of Victim Services, 1993, p. 1567).

The Gas Mart where Jane was killed had signs plastered all over the windows. Automobile and pedestrian traffic was unable to see into the store or to observe any crime in progress, permitting the murderer to commit his crime in relative privacy. His only concern was that of a random shopper entering the store before he left. Most small convenience stores have large expanses of windows and if they do not, they should. These windows should be kept open to the street with all signs removed. A would-be thief is more likely to avoid a well-lit, windowed store because a crime in progress can be more easily spotted by a passer-by or police patrol car in the area.

Jane was wearing street clothes when the crime in her store took place. All store personnel should wear distinctive smocks or uniforms. Many stores engage in this practice for image purposes, but it is also a sound safety practice. In the event the police are summoned to a crime in progress, it will be easier for them to distinguish the perpetrator from the victim.

In spite of the above precautions, if a robbery does occur, store personnel should be instructed to fully cooperate with the robber. The silent alarm should be activated only if it can be done so without giving notice to the assailant. Instruct employees to only speak in direct response to the perpetrator's questions and not to volunteer additional information. They should be as observant as possible and attempt to avoid confrontation (Butterworth, 1993).

Once the robber leaves the store, employees should immediately lock the store and call the police. They should ask witnesses to stay until police arrive and to not discuss the incident until that time. Everyone should avoid touching any surfaces that the perpetrator may have touched (Butterworth, 1993).

Anticipating and preparing for events such as robbery should be an important part of any employer's plans. This is especially true for night-time retailers. The costs of formalizing and executing these plans is minimal compared to the potential saving of the lives and well-being of employees. Although legislation has not forced action in every state, it is forthcoming.

To avoid costly hold-ups, potential litigation, and to prepare for inevitable regulation, it makes sense to take these relatively inexpensive steps to protect your business investment and employees against robbery and violence.

WORKPLACE EPIDEMIC

Greg Couls, Michael Konz, Ron McTasney, and John Scully are four of more than 1,300 victims who died in 1994 from America's newest epidemic ("Gunmen Kills," 1994, p. A-7). This epidemic cannot be cured by antibiotics, chemotherapy, radiation treatments, AZT, or surgery. America's finest medical researchers *will not* develop a vaccine to cure this epidemic. The manifestation of this epidemic frequently results in gruesome, emotionally scarring events with deadly results. It's called workplace violence, and it is laying siege of America's workplace and workforce.

"Workplace violence is the new poison of corporate America" (Dunkel, 1994, p. 40), says Dennis Johnson, a clinical psychologist and President of Behavior Analysts & Consultants, a Stuart, Florida--based management firm that closely tracks workplace violence. "It is not just a reflection of a violent society, but of that violent society interacting with workplace dynamics that are significantly changed from 10 or 15 years ago" (Dunkel, 1994, p. 40).

Background

Recently, an owner of a pawn shop, a convenience store clerk, a psychologist, two sanitation managers, a tavern owner, a fisherman, a cook, two cab drivers, a co-owner of a furniture store, a restaurant manager, a maintenance supervisor, a video store owner, and a postal carrier were all victims of workplace homicide in the same week. The National Institute for Occupational Safety and Health (NIOSH) estimates fifteen people are murdered at work each week in the United States.

During the last decade, NIOSH documented nearly 7,600 homicides in United States workplaces. Workplace homicide was the principal cause of occupational death for women and the third principal cause of death for all workers in the 1980s ("NIOSH Urges," 1993).

In 1994, workplace homicide accounted for 20 percent or 1,318 of the 6,588 workplace fatalities according to statistics recently released by the Bureau of Labor Statistics. This was second only to deaths resulting from car and truck accidents which accounted for 42 percent of work-related fatalities. Homicide was again the number one cause of death of working women, paralleling NIOSH's study of the 1980s ("6, 271 Job-related," 1994).

A recent Satellite Video Conference presented by The George Washington National Satellite Network (1994) on Workplace Violence presented the following chilling statistics. These statistics summarize the vastness of the workplace violence epidemic:

- From 1987-1992, 1 in 6 non-fatal assaults occurred in the workplace.
- Occupational homicide has been the leading cause of death by injury in the workplace for women since 1980.
- In 1992 and 1993, occupational homicide was the number 2 cause of workplace death for men.
- Fifteen percent of U. S. workers have been physically attacked at work.
- One in four U. S. workers will be attacked, threatened, or harassed during their work career.

High-Risk Workplaces

All workplaces are vulnerable to violence: family-owned businesses, government, major corporations, manufacturing, military, non-profit, private, public, retail, service, and other small businesses (The George

Washington National Satellite Network, 1994). Among U. S. workplaces, retail trades have had the highest number of occupational homicides during NIOSH's study period. Service industries were a distant second. However, these two types of workplaces accounted for 54 percent of all workplace homicides during the NIOSH study period ("Preventing Homicide," 1993).

According to 1992 Bureau of Labor Statistics data, retail trades accounted for 48 percent of workplace homicides, service industries 19 percent, and transportation/public utility 12 percent. These three industry groups accounted for 79 percent of the workplace homicides in 1992. Refer to Table 12-1 for further details (Novicio, 1994).

High-Risk Occupations

The occupation with the highest rate of occupational homicide was taxicab drivers/chauffeurs. Other high-risk occupations included law enforcement officers, hotel clerks, and gas station workers. Refer to Table 12-2 for further details ("Preventing Homicide," 1993).

Table 12-1. 1992 Homicides by type of workplace

WORKPLACE	PERCENT OF DEATHS
Retail Trades	48 %
Service Industries	19 %
Transportation/Public Utilities	12 %
Public Administration	7 %
Finance/Insurance	4 %
Manufacturing	3 %
Construction	2 %
Wholesale Trade	2 %
Agriculture	2 %
Miscellaneous	1 %

Table 12-2. High-risk occupations

OCCUPATION	DEATH RATE PER 100,000
Taxicab Drivers/Chauffeurs	15.1
Law Enforcement Officers	9.3
Hotel Clerks	5.1
Gas Station Workers	4.5
Security Guards	3.6
Stock Handlers/Baggers	3.1
Store Owners/Managers	2.8
Bartenders	2.1
Avg. for all Workers	0.7

Cost to Business

Workplace violence has a staggering impact on the emotional and financial health of an organization. Some of the direct and in-direct costs include direct legal and medical costs, out-of-court settlements, death benefits, employee assistance programs, security-related services and products, business disruption, public and media relations, diverted resources, emotional scarring, extended litigation, fear, lost personnel, name/product tainting, facility repairs, turnover, and wasted knee-jerk reaction costs.

One incident of workplace violence can decimate a small business and ravage larger businesses. A recent survey completed by the Workplace Violence Research Institute indicates that the costs generally associated with workplace violence are conservative at best. The survey revealed the following cost figures:

■ Workplace violence has cost business over $36 billion since 1990.

■ A single incident of workplace violence costs about $1 million.

■ Victims of workplace assault lost an average of 5 work days. This equates to approximately $69 million in lost wages per annum.

■ Rape/sexual assault of women is the number one source of security liability claims. The average settlement of these claims is $545,000; the average verdict award is $3.5 million ("Survey Reveals," 1995).

VICTIMIZATION OF THE AMERICAN WORKFORCE

Severity of Victimization

Approximately one million employees become victims of violent crime while performing their jobs each year. These attacks account for 15 percent of the over 6.5 million acts of violence sustained by United States residents age 12 or older. Additionally, over two million personal thefts and 200,000 car thefts occur during working hours each year (Bachman, 1994).

The National Crime Victimization Survey conducted from 1987–1992, highlighted several key points:

■ Criminal attacks in the workplace cause approximately 500,000 employees to lose over 1.7 million workdays annually.

■ Men were more likely than women to experience a violent crime. However, women were as likely as men to become victims of theft while working.

■ Over 30% of victims faced an armed aggressor.

■ It is estimated that over half of all workplace victimizations are not reported to the police (Bachman, 1994, p. 1).

Upon questioning victims on why they did not report attacks, 40 percent stated they believed the event was minor or a private matter. An additional 27 percent did not report because they reported the incident internally to company representatives. These representatives included security officers, human resource department personnel, or their manager (Bachman, 1994).

In October of 1993, Northwestern National Life Insurance released the results of a telephone survey they conducted of 600 American workers entitled "Fear and Violence in the Workplace." The conclusions were macabre enough to make individuals shop for body armor that goes well with their favorite business suit. It estimated that more than 2.2 million Americans were attacked at work during 1992 and another 6.3 million threatened. Moreover, 15 percent of the survey participants said they had been physically assaulted on the job at least once ("Fear and Violence").

Profile of Victims

The victims of workplace violence can be anybody and no one is immune. A recent Satellite Video Conference presented by The George Washington National Satellite Network presented the following victim profile:

- Approximately 83 percent of workplace homicide victims were men.
- Workplace homicide accounts for 33 percent of working 13 to 17-year-olds' deaths.
- Workplace homicide victims were 75 percent white, 19 percent black, and 6 percent other.
- Five percent of the women victimized at work were attacked by a husband, boyfriend, or "ex."
- An estimated one in four women will be sexually assaulted at work.
- Assault is the leading cause of injury for women ages 16 to 55.
- Six of ten incidents of non-fatal violence occurred in private companies.
- Thirty percent of non-fatal assault victims were government employees, although they comprise only 18 percent of the U.S. workforce ("Preventing Homocide," 1994).

FORMULA FOR FAILURE

The Ingredients List

Fear and violence in the workplace can be ascribed to various factors:

- Robbery—predominant motive in occupational homicides
- Technological innovations
- Prevalence of violence in society
- Diversity and change in the workplace
- Shifts of responsibility at home
- Downsizing (rightsizing, wrongsizing, re-engineering) ("Preventing Homocide," 1993)

Dennis Johnson has examined the hows and whys of hundreds of workplace violence cases. The "dynamics" Johnson and other experts cite for the increase in workplace violence radiate from the following factors:

- A hyperactive, ever-evolving populace
- An acutely over stressed society
- An overabundance of guns
- Fractured families
- Numerous self-heralded victims produced by what the Art Critic Robert Hughes calls "the culture of complaint"
- Workplace diversity and the inability to adjust to it (Dunkel, 1994, p. 40)

The factors from the two lists above create an incredibly complex recipe. Because of the detail required to adequately address each of these factors, the focus will be on one of the more prevalent areas which has had considerable impact on the American workplace—downsizing.

Blending the Ingredients

Downsizing, re-engineering, rightsizing, restructuring, and wrongsizing have become five of the most common terms in today's business nomenclature. Many Americans have fallen victim to this phenomena of the 1990s.

The American workplace has changed. American business is struggling to make the transition to a high-tech, global economy. Rarely does a week go by in which a major corporation does not make an announcement that a number of jobs are being eliminated.

Aggressive corporate re-engineering is creating a multitude of psychologically crippled workers who have seen their job prospects and self-esteem vanish. In the 1970s, notes Johnson, 80 percent of all displaced employees could find a comparable replacement job. In the 1990s, that total has been hovering at less than 25 percent (Dunkel, 1994).

Some companies have targeted positions that can be eliminated by attrition. This can be a golden opportunity for those close to making the retirement decision. However, for individuals on the fringe, it is potentially devastating. These individuals while given a good separation package (severance pay, continued benefits, outplacement services, etc.) may simply not be ready, nor able to afford, retirement.

Manufacturing jobs, which pay equivalent wages of the lost job, have disappeared. The alternative, America's fastest growth area, is the retail industry. There is one major problem. Most of the available jobs pay minimum wage or slightly higher. For the typical family with a mortgage, car and other debt payments, minimum wage is not enough.

Baking the Ingredients

A highly stressed workplace is the most susceptible to violence. The following factors can add considerable fuel to the fire:

- Management does not talk with or delegate control to employees.
- Employee work is fast paced and performed in poor environmental conditions.
- Overtime is frequent and mandatory.
- Employee benefits have recently been cut (Joyner and McDade, 1994, p. A-8).

Couple workplace downsizing with a combination of any or all of the above and a stressful situation has been created. As stress from the workplace and family builds, the displaced worker who may already be on the edge becomes more desperate and isolated. This combination can become lethal. It is a time when an act of violence could occur.

RECOGNIZING THE POTENTIAL AGGRESSOR

Before acts of violence are discussed, it is important to have an understanding of who may be a potential aggressor. Aggressors can be anybody: clients, competitors, criminals, current and former employees, current and former relationship partners, customers, drug addicts, gangs, or terrorists. Whoever the aggressor may be, there are usually a number of red flags present that should not be ignored.

Disgruntled Employee Red Flags

Supervisors and managers need to pay attention to employees who exhibit marked changes in their demeanor and performance. Employees who are normally passive and quiet who suddenly become loud, boisterous, and disgruntled should be closely observed. This may be the beginning of a potentially volatile situation.

During a recent Satellite Video Conference on Workplace Violence presented by The George Washington National Satellite Network (1994), FBI Special Agent Eugene A. Rugala tendered the following list of red flags that should be closely monitored:

- History of violent behavior
- Obsession with weapons, compulsive reading, and collecting gun magazines
- Carrying a concealed weapon
- Direct or veiled threats
- Intimidation or instilling fear in others
- Obsessive involvement with job
- Loner
- Unwanted romantic interest in co-worker
- Paranoid behavior
- Unaccepting of criticism
- Holds a grudge

- Recent family, financial and/or personal problems
- Undue interest in publicized violent events
- Tests limits of acceptable behavior
- Stress in the workplace such as layoffs, reduction in forces and labor disputes
- Any extreme changes in behavior or stated beliefs (Rugala, 1994)

Special Agent Rugala emphasized one key point: "an employee can manifest one or all of these traits and never act out violently! This is where the importance of having solid employee/supervisor relationship cannot be underscored!" (Rugala, 1994.)

Profile of Workplace Aggressors

A question many people ask is "What does the workplace aggressor look like?" This question cannot be easily answered. The following profile presented by Special Agent Rugala (1994) cannot be cast in concrete as research in this area is still in its infancy. However, the following factors are tendencies that have been seen in numerous cases:

- Thirty to 50 years old
- Little or no criminal record
- Paranoid, aggressive behavior
- Troublemaker, unstable work history
- Substance abuser
- Owns several guns
- Family and marital problems (Rugala, 1994)

In cases where an individual is being observed for potential violent behavior traits, these factors form a foundation. It is important not to focus too much attention on the above tendencies and become blind to an employee who is showing signs of violent behavior but does not fit the profile.

Employee Disenchantment

Why do good employees suddenly change? Why do bad things happen to good people? What causes top performers to suddenly bottom out? What causes these employees to drop off the deep end? Something must happen to cause these employees to snap.

Employee disenchantment can be a major source of stress and, through time, this stress can build to the point of the person "snapping" according to Special Agent Rugala. There are numerous causes of employee disenchantment. The following list highlights causes of employee disenchantment:

- Confusion
- Lack of trust
- Not being listened to
- No time to solve problems
- Office politics
- Someone solving problems for you
- Not knowing whether you are succeeding
- Indiscriminate rules
- Boss takes credit for others ideas and work
- Believing you cannot make a difference
- Meaningless job (Rugala, 1994)

Working under the above conditions certainly adds stress. As was discussed previously, this continual stress can build to the point where the individual cannot take it anymore and finally breaks. Hopefully, this breaking point will not result in an act of violence.

ACTS OF VIOLENCE

According to J. Branch Walton (1994), former Director of Corporate Security for Cummins Engine, acts of violence generally fall into two categories: revenge and violence.

Revenge

Revenge comes in a variety of forms. Carefully plotted revenge can destroy a company. The following are different types of revenge:

- Product tampering
- Rumor spreading
- Theft
- Theft of secrets
- Violence
- Stalking
- Kidnaping executives
- Threats
- Harassment

The following are some recent examples of revenge:

- The Tylenol tampering incident in the 1970s had a profound effect on Johnson & Johnson as well as every other manufacturer of consumer medical products.

- Encyclopedia Britannica experienced a serious product tampering incident when a discharged employee changed all references to Jesus to Allah. Thousands of copies were printed before the change was discovered.

- A terminated employee on his way out of the company's headquarters building lobby gouged several priceless Van Gogh paintings destroying them.

- A discharged telephone company employee tapped the telephones of several prominent citizens in the community and then attempted to extort money from them to keep the information he had acquired quiet.

- An employee of a fast food restaurant who was fired changed the slogan on the cash register receipt, insulting numerous customers.

These are only a few examples of revenge. As can be seen by these examples, significant damage can be caused to a company's image by a malicious employee wanting to pay back the company for being discharged.

Violence

Unfortunately, some individuals resort to violent acts. These acts include

- Homicide
- Rape
- Assault
- Kidnaping
- Threats (Branch Walton, personal communication, June 23, 1994)

Each of these violent acts is well documented. Acts of violence are growing at an alarming rate. Then NIOSH Acting Director Richard A. Lemen stated, "The time has come to take action to prevent these tragic crimes. We may not have all the answers at this point, but we do know that there are protective measures that may help end these senseless deaths ("NIOSH Urges," 1993, p. 1).

DEFUSING A TIMEBOMB: THE VIOLENT EMPLOYEE

While there is no current cure for workplace violence, there are proactive, preventive steps that can be implemented. These steps form the basis for a program that will expand the company's human resources, safety, and security programs.

According to Mr. Walton (1994), the following seven preventive steps provide a realistic approach to defusing workplace violence before it occurs. These seven preventive steps are

- Establish a clear nonharassment policy.
- Perform pre-employment screening.
- Establish a drug testing program.

- Conduct employee and management training in stress management and communications.
- Do crisis management planning.
- Establish proper security measures.
- Foster a working liaison with local law enforcement (Branch Walton, personal communication, June 23, 1994).

Nonharassment Policy

Senior management must make it clear to its employees that harassment and threats will not be tolerated. Garry Mathiason, a San Francisco lawyer who specializes in workplace liability law stated, "If you have a policy or plan in place, then what is tolerated and what is not become part of the culture." (Dunkel, 1994, p. 70).

This is especially important with verbal threats. Company policy must be firm that verbal threats will not be tolerated. Mathiason goes on to say, "If you doubt me, (ask yourself) when was the last time you made a joke going through an airport metal detector?" (Dunkel, 1994, p. 70).

Pre-employment Screening

The significance of this tool can not be understated. How involved your organization gets with pre-employment screening will be a function of the nature of your company's business, culture, and the sensitivity of the position being filled.

Before any pre-employment screening is done, consult with your company's legal counsel and discuss the proposed pre-employment screening to be conducted. This is vitally important and can save your organization significant expense should a pre-employment screening tool be considered to discriminate in your state (Lynn C. Outwater, personal communication, February 15, 1995).

Target Stores in Oakland, California, asked prospective security guards to take the Rodgers Condensed CPIMMPI written prescreen (a psychological screening tool). An attorney for the plaintiffs, in a class-action lawsuit, hinted that a number of the questions on the long-used exam were "extremely invasive on matters of sexuality, religion, bodily functions, and the like" (Albrecht and Mantell, 1994, p. 50). Under the settlement of agreement, Target Stores paid $1.3 million to 2,500 awardees and agreed to ban the test in its 113 California stores.

What pre-employment screening tools should be used? At a minimum the following five are a must:

- Work history verification
- Military history verification
- Credit history
- Driving record check
- Criminal history check

These five areas can provide significant information about an individual. Several red flags would be unexplained gaps of greater than 30 days in work history; scant details of disciplinary matters in military service; dire financial circumstances (high debt, home foreclosure, etc.) in credit history and numerous accidents or tickets on a driving record. Obviously, you must look at these from a big-picture perspective, one negative report may not be too bad, but taken as a whole may be significant (Branch Walton, personal communication, February 15, 1995).

Drug Testing

The company should establish a drug testing program. These programs can be quite effective when properly administered. As with the pre-employment screening, company legal counsel and personnel experienced with Employee Assistance Programs and drug testing in the workplace should be consulted before starting this program (Lynn C. Outwater, personal communication, February 15, 1995).

Once the foundation of the drug testing program has been established, drug testing should be conducted in three areas: pre-employment, random and for cause or reasonable suspicion. Each of these areas is quite specific and requires the guidance of professionals in ensuring they are properly administered. Consult with these individuals for further details in establishing and administering your company's program (Lynn C. Outwater, personal communication, February 15, 1995).

Employee and Management Training

Employee and management training provided in several areas is critical to successfully handling a potentially explosive situation. These areas include stress management, conflict resolution, negotiating, and communications. The training must be tailored to the company's philosophies and culture.

Training should be presented at two levels. First at the employee level to establish a baseline—a common foundation. Second, at supervisory and managerial levels to instill a consistent manner in which various situations will be handled. Additionally, the training should include role playing to provide participants the opportunity to experience a variety of situations in a controlled learning atmosphere (Branch Walton, personal communication, June 23, 1994).

Crisis Management Planning

Planning for an incident involving workplace violence should be no different than planning for a fire or chemical spill. Pre-planning for a workplace violence incident is critical. Having a plan of action in place will provide structure to an otherwise chaotic situation.

What should be in a crisis management plan for handling a workplace violence incident? In their book *Ticking Bombs: Defusing Workplace Violence in the Workplace (1994),* Steve Albrecht and Michael Mantell provide the following list:

- Telephone numbers of local law enforcement and emergency services
- Notification of key company personnel should an incident occur. This list must include the company's legal counsel
- Procedure for protecting the scene of a workplace violence incident for investigators
- Procedure for checking the integrity of your company's data and computer systems
- Arrangements and retainer contracts for on-scene employee counseling immediately following a significant event (24 hour coverage)

- Training of senior site management in dealing with the media
- Designating an official company spokesperson and media relations procedure
- Procedure for providing grief- and trauma-recovery time for victims and related witnesses, bystanders, and employees
- Procedure for company sanctions and/or punishment for the instigators in lesser cases
- Arrangements with company Employee Assistance Program and psychological counseling programs to help employees cope with post- incident stress management
- Procedure for handling victim(s) family notification(s) and the company's aid package
- Procedure for handling clean up and scene restoration (Albrecht, 1994, p. 221-242)

While the steps above deal specifically with handling an incident of workplace violence, another important procedure must be developed. This procedure is a termination procedure.

The termination procedure should include details on how to handle the termination process. This procedure should include

- Handling the individual's personal effects
- Preventing facility access
- Collecting company property (credit cards, keys, access cards, identification badges, proprietary information, etc.)
- Handling outstanding expense reports
- Handling severance, vacation pay, and other continued benefits
- Handling difficult employee terminations (Branch Walton, personal communication, June 23, 1994)

These steps are not all inclusive and may not apply in all situations. They must be tailored to each company's structure and uniqueness.

Proper Security Measures

Sound security measures begin with the basics, controlling access to the facility. Each entrance to a building should have some type of access control device. These devices can be as simple as keyed entrances and as complex as magnetic card readers which permit access to authorized areas only.

Another access control measure is controlling visitor entrance to the facility. The company should have a specific procedure for handling visitors. Visitors should be required to sign in at specified points and be accompanied by a company representative at all times. They should never be given the freedom to roam at will within a facility.

One final means of controlling access to a facility is to ensure devices installed to control facility access are not compromised. This includes purposefully leaving doors open, revealing access codes, duplicating keys, and leaving fence gates unlocked. *Access control is only as good as management's commitment to facility security management.*

Another basic security measure is to issue photo-identification cards. These identification cards should have a color photograph. In addition, the following information should be on the identification badge:

- Employee's name, signature, and social security number
- Employee identification badge serial number
- The signature of the person authorized to authenticate and issue identification badges

There are numerous other security measures which can be undertaken. The key is to customize the measures necessary for your company's needs. Finally, whether your company takes a low-profile approach (access control, identification badges) to security or a highly visible approach (closed circuit television cameras, private security officers), a facility's security management program should be given the same management attention as safety, operations, sales, and marketing (Branch Walton, personal communication, June 23, 1995).

Liaison with Local Law Enforcement

This is an area whose importance can not be understated. Having an open line of communication with local law enforcement is of untold value. During a crisis situation, it is nice to know the key personnel in the local organization and their commitment to assisting your company.

Another area where local law enforcement can be invaluable is helping your company understand local restraining laws and restraining orders. These may have an impact on your company if employees have a restraining order in place against a spouse or "ex" (Branch Walton, personal communication, June 23, 1995).

The old adage, "An ounce of prevention is worth a pound of cure" can be applied to successfully combating workplace violence. This seven-step model, when applied, forms that "ounce of prevention." This plan does not guarantee workplace violence will not occur. However, it will better prepare your organization should an event occur it will and protect you, your workplace, and employees from America's next epidemic.

CONCLUSION

This chapter reviewed the occupational epidemic sweeping the United States that is commonly referred to as workplace violence. It is a problem that many safety professionals are beginning to address in an attempt to reduce the risk of serious injury to thousands of working men and women across America. The workplace aggressor can be anyone from a disgruntled employee to a psychologically disturbed competitor. Workplace violence can be as traumatic as a "madman" shooting an ex-boss and co-workers or as insidious as product tampering. Like any other loss in the occupational environment, the prevention of workplace violence requires a comprehensive approach that includes everything from pre-employment screening to crisis management planning. No workplace is immune from this epidemic. With workplace homicide being the number one cause of occupational death for women and the third leading cause of death for all workers, safety professions must expand their lengthy list of responsibilities to include workplace security.

Questions

1. What is workplace violence? How far-reaching is this problem in the United States?

2. Why is knowledge of workplace violence important to a safety professional?

3. What are some of the factors that contribute to workplace violence?

4. List and describe some of the "red flags" that could indicate that a disgruntled employee might be a candidate for workplace violence.

5. What are some of the acts of violence associated with workplace violence?

6. List and briefly describe the components of a comprehensive workplace violence program.

References

Albrecht, S., and Mantell, M. 1994. *Ticking Bombs: Defusing Violence in the Workplace*. Burr Ridge, IL: Irwin Professional Publishing.

Bachman, R. 1994, July. Violence and theft in the workplace. *Crime Data Brief*, pp. 1-4.

Bell, C. A. 1991, June. Female homicides in US workplaces. *American Journal of Public Health*, pp. 729 - 732.

Butterworth, R. A. 1994. *Convenience Business Security Act: Robbery Deterrence and Safety Training Guidelines*. Tallahassee, FL: Office of the Attorney General.

Carder, B. 1994, February. Quality theory and the measurement of safety systems. *Professional Safety, 39.* 23-28.

Catania, A. C. 1968. *Contemporary research in operant behavior.* Glenview, IL: Scott, Foresman and Company.

Daniels, A. C. 1989. *Performance Management: Improving Quality Productivity Through Positive Reinforcement.* Tucker, Georgia: Performance Management Publications.

Dunkel, T. 1994, August. Danger zone: Your office. *Working Woman, 19,* 39 - 71.

Florida Department of Legal Affairs, Division of Victim Services. 1993. *Additional Statement to the Secretary of State Rule* (Chapter 2A - 5). Tallahassee, FL: Legal Document Reproduction.

Gunman kills 2, wounds 2 at Ford union meeting. 1994, August, 14. *Buffalo News*, p. A7.

Homicide in US workplaces. 1992, September. *U.S. Department of Health and Human Services,* 1-7.

Johnson, D. L., Kiehlbauch, J. B. and Kinney, J. A. 1994, February. Break the cycle of violence. *Security Management, 38,* 24-28.

Johnson, D. L. and Kinney, K. A. 1993. *Breaking Point.* Chicago, IL: National Safe Workplace Institute.

Jorma, S. 1994, May. When does behavior modification prevent accidents? *Leadership & Organization Development Journal, 15.*

Joyner, T. and McDade, S. 1994, March 31. Workplace violence blamed on stress. *Niagara Gazette.* p. A8.

NIOSH urges immediate action to prevent workplace homicide. 1993, October 25. *NIOSH Update.*

Northwestern National Life Employee Benefits Division. 1993. *Fear and Violence in the Workplace.* Minneapolis, MN.

Novicio, J. 1994, February 11. On-the-Job Homicide Increases. *Niagara Gazette*, p. A6.

Preventing homicide in the workplace. 1993, September. *NIOSH Alert*: U. S. Department of Health and Human Services.

Rugala, E. A. 1994. *Recognizing the Potential Aggressor.* Address presented at The George Washington National Satellite Network Video Conference: Washington, DC.

6,271 job-related deaths counted in 1993; Vehicle crashes, homicides lead all causes. 1994, July 27. *Niagara Gazette*, p. A6.

Sprouse, M. 1992. *Sabotaging the American Workplace.* San Francisco, CA: Pressure Drop Press.

Survey reveals the cost of workplace violence to employers. 1995, May 15. *OSHA General Industry News*. Santa Monica, CA: Merritt Publishing.

Tarasuk, V., and Eakin, J. 1994, May/June. Adding insult to injury. *OH&S Canada,10*, 44-47.

Toscano, G. and Windau, J. 1994, October. The changing character of fatal work injuries. *Monthly Labor Review, 117*, pp. 17-18.

Chapter 13

Hazardous Materials

CHAPTER OBJECTIVES

After completing this chapter, you will be able to

- Outline the basic recent environmental legal history.
- Understand the Resource Conservation and Recovery Act (RCRA).
- Identify the types of hazardous waste.
- Determine hazardous waste generator status.
- Know how to obtain an EPA identification number.
- Understand the Comprehensive Environmental Response and Compensation Liability Act (CERCLA).
- Understand the Superfund Amendments and Reauthorization Act (SARA).
- Explain the Hazard Communications Standard (HAZ COM).

CASE STUDY

Two employees of a North Carolina company were helping to unload a truck on a warm sunny day when they inadvertently punctured a container of an unknown liquid chemical. A quick check of the contents revealed that they had spilled a concentrated pesticide, intended to be mixed with water before application. Their first response was to grab a garden hose and use the water pressure to push it to one side. In a short time they began to feel faint and short of breath, so they decided to ask for help. Since no one at the facility had specialized training to deal with the spill, the manager on duty called the local fire department.

When the fire department showed up, it soon became evident that their personnel also lacked training to deal with spills of hazardous chemicals. They formulated a plan to wash the remainder into the storm drain and use a vacuum truck to intercept the waste as it flowed through the system. A few minutes after the flushing began, the crew at the spill site received a call from the group at the truck asking how soon the flushing would begin. As it turns out, the truck was stationed at the wrong storm sewer opening and it missed the contents altogether. By the time all the mistakes were remedied, clean-up costs amounted to over $50,000.

INTRODUCTION

The increase in environmental regulation is probably the most significant legal development in America in the last 50 years. Safety practitioners do not need a detailed knowledge of the fine points of these laws and regulations. They do, however, need a basic understanding of the structure, scope, and framework of these laws in order to effectively measure and evaluate the nature of the compliance task faced.

Many of the environmental problems faced by the safety practitioner are covered or touched on by the Resource Conservation and Recovery Act (RCRA). When this act was passed in 1976, Congress was attempting to complete a series of legislative acts designed to protect the environment. A brief overview of some of the legislation that lead to the passage of RCRA will be useful.

The Hazardous Materials Transportation Act (HMTA) was passed in 1975 to give the Department of Transportation (DOT) the authority to regulate the transportation of hazardous materials by air, waterways, rail, or highway. To avoid exposing large population areas to significant risks from exposure to hazardous materials, the DOT was given authority to establish routes for hazardous materials shipping.

That same year, Congress passed the Toxic Substances Control Act (TSCA) which gave broad powers to the Environmental Protection Agency (EPA) to regulate the production and use of potentially hazardous chemicals and to assure that new chemicals did not pose unreasonable hazards. Under TSCA, EPA may require a company to submit test data on specified chemicals. Anyone who intends to manufacture or import a chemical not listed on the TSCA Chemical Substance Inventory must submit a premanufacture notice (PMN) at least 90 days in advance of the manufacture or import of a new chemical. EPA may then prohibit the manufacture, process, or distribution of the chemical. It may also place limits on the chemical's production or ban its production.

After the passage of HMTA and TSCA, Congress put the finishing touches on its hazardous material legislation with the passage of RCRA. RCRA controls the chemicals at their source during transportation and disposal. The RCRA requirements often fall within the scope of the safety practitioner's responsibilities, not only because of the natural relationship that management perceives between environmental and safety responsibilities, but also because of subsequent legislation which will be discussed later.

BACKGROUND

The Resource Conservation and Recovery Act (RCRA) was passed by Congress to direct the United States Environmental Protection Agency (EPA) to implement a program that would "protect human health and the environment from improper hazardous waste management" (US EPA, 1991b). RCRA can be administered by the EPA or under state EPA plans. State-administered EPA programs must meet and/or exceed the standards set by the federal EPA. The first set of RCRA regulations was published in the *Code of Federal Regulations* under Title 40 which deals with "Protection of the Environment."

Early regulations focused on large companies producing more than 1,000 kilograms of hazardous waste per month. Subsequent amendments directed the EPA to also include producers of hazardous waste between 100 kilograms and 1,000 kilograms per month.

WHAT IS A HAZARDOUS WASTE?

Basically, a waste is something unwanted, usually a solid, liquid, or a contained gas. It is the responsibility of the generator to determine if the waste is nonhazardous, hazardous, or acutely hazardous. The EPA classifies a waste as hazardous if, through improper handling, it can cause injury or death or can damage or pollute the environment (US EPA,

1991). To determine if a waste is hazardous, it may be necessary to refer to the *Code of Federal Regulations* at 40 CFR Part 261, "Identification and Listing of Hazardous Waste." In this section, a waste is considered hazardous if it is (1) a *listed waste* or (2) a *characteristic waste*.

The regulations include four lists identified by the letters F, K, P, and U that together describe about 400 hazardous wastes (US EPA, 1991). The F-list includes hazardous waste from non-specific sources or generic wastestreams, whereas the K-list covers hazardous wastes from specific sources. F-list wastes are more general in nature and K-list wastes are generated by a specific manufacturing process. Those on the P-list are considered acutely toxic hazardous wastes. The U-list identifies commercial chemical products, chemical intermediates, and off-specification chemical products. Some states may include additional lists in this section, such as Kentucky's N-list which covers nerve and blister agents (Chapter 31, 1992). Wastes that appear on any of these sections are referred to as listed wastes."

If a waste cannot be found on one of the RCRA lists, this does not mean it is not hazardous. Unlisted waste must still be tested to determine if it has certain properties or characteristics that render it hazardous (US EPA, 1991). A waste is hazardous if it has one or more of the following properties: ignitability, corrosivity, reactivity, or toxicity.

Ignitable wastes have a flash point lower that 140 degrees Fahrenheit. *Corrosive* wastes are those that are acidic (at or below a pH of 2) or caustic (at or above 12.5 pH). *Reactive* wastes produce violent results when mixed with water, air, or other chemicals. Results include explosions and the generation of toxic gases. *Toxic characteristic* wastes contain specified percentages of specific metals, pesticides, or organic chemicals. They are discovered by testing the wastestream using the Toxicity Characteristic Leaching Procedure commonly known as a TCLP. This test is used to determine the amount of chemical that would leach out into the ground water under specific conditions. Wastes that exhibit any one of the above characteristics are referred to as *characteristic wastes* or *TCLP wastes*. They are coded as *D wastes*.

DETERMINING GENERATOR STATUS

Once it has been determined that the waste produced by a plant is hazardous, the facility becomes a hazardous waste generator. To determine the generator status of a facility, the total hazardous waste it generates per calendar month must be determined. The correct generator status is important because different regulations apply to each different status.

A facility is a *limited-quantity generator* if it produces less than 100 kilograms of hazardous waste per month. If, however, a limited-quantity generator stores more than 1,000 kilograms of hazardous waste on-site at any time, it becomes a small-quantity generator. A limited-quantity generator is not required to obtain an EPA identification number, but most licensed waste haulers will require an EPA identification number to ship the waste off-site.

A facility is a *small-quantity generator* if it produces between 100 kilograms and 1000 kilograms of hazardous waste per month. Small-quantity generators are required to obtain an EPA identification number. They can accumulate up to 6,000 kilograms of hazardous waste in any 180-day period. If, however, the hazardous waste is to be transported more than 200 miles, the generator is allowed to accumulate hazardous waste up to 270 days.

A *large-quantity generator* is a facility that produces 1,000 kilograms of hazardous waste or one kilogram of acutely hazardous waste per month. Large-quantity generators are allowed to accumulate hazardous waste for 90 days. If the hazardous waste must be transported more than 200 miles, the generator is allowed to accumulate hazardous waste for 180 days (US EPA, 1991).

Only limited-quantity generators are exempt from most RCRA regulations. Other generators must have an EPA identification number. Some states may have more stringent regulations than those of the federal government. The practitioner should check with the state EPA if there are any questions about a given facility.

OBTAINING AN EPA IDENTIFICATION NUMBER

To obtain an EPA identification number, a Notification of Hazardous Waste Activity (Form 8700-12) must be requested from the USEPA or state EPA. If the facility is a Treatment Storage and Disposal (TSD) facility, additional forms must be completed. Once the EPA has reviewed the form, a 12-character EPA ID number will be assigned to the facility (Kaufman, 1990). This form must be modified each time additional waste streams are generated, deleted, or generator status changes.

MANAGING HAZARDOUS WASTE ON-SITE

Hazardous waste can be stored on-site in containers or tanks (Chapter 31, 1992). Proper storage during the accumulation period is necessary to prevent accidents and/or spills. Some of the basic requirements for temporary storage include the label "Hazardous Waste", proper identification of the waste, and the accumulation start dates for each container or tank (Chapter 31, 1992). Containers or tanks must be kept closed except when filled or emptied, and operational log books must be kept to record any activity. They must be inspected weekly for leaks. Safety equipment must also be inspected weekly. Inspection reports and a written emergency plan must be kept on file. Incompatible chemicals must be segregated and three feet of aisle space must be maintained in the accumulation area. Signs must be posted identifying the area as a hazardous materials storage area.

CERCLA

Once RCRA was passed in 1976, Congress no doubt believed that most of the environmental problems had been addressed. It wasn't until the discovery of Love Canal and other similar problems in the late 1970s that Congress realized there was a significant hole in its plan.

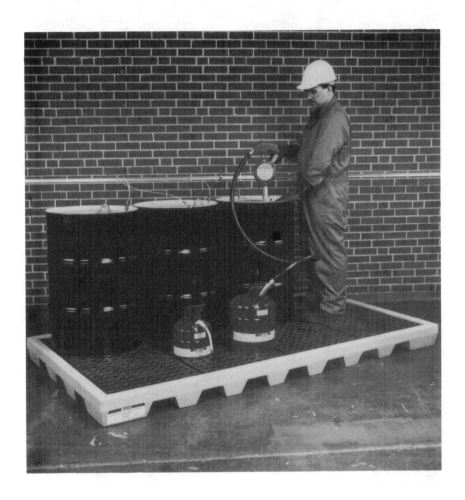

Figure 13-1. Example of a hazardous material accumulation/dispensing center with spill containment. (Photograph courtesy of Justrite Mfg. Co. L.L.C., Des Plaines, IL)

Figure 13-2. Example of a single drum hazardous material collection station. (Photograph courtesy of Justrite Mfg Co. L.L.C. (Des Plaines, IL)

Figure 13-3. Example of a hazardous material four-drum pallet with spill containment. (Photograph courtesy of Justrite Mfg. Co. L.L.C.; De Plaines, IL)

Figure 13-4. (a) Example of hazardous material perforated absorption mat (b) Example of spill containment "sock." (Photograph courtesy of New Pig Corporation, Tipton, PA)

Love Canal was built near Niagara Falls, New York, in the 1800s to link waterways. Although it was never completed, the canal, in the form of a half-mile ditch, remained until it was purchased by the Hooker Chemical Company in the early part of the twentieth century, as a dump site for hazardous waste. Once full, the canal was covered and eventually wound up as the building site for a school with residential properties nearby.

Once this and other abandoned sites were exposed as significant threats to the environment and residents, Congress realized it had not addressed the problems associated with existing waste sites. Thus, in 1980, the Comprehensive Environmental Response and Compensation Liability Act (CERCLA) was passed. CERCLA or Superfund as its sometimes called, gives the government authority to pursue potentially responsible parties (PRPs) for past and future cleanup costs at abandoned waste dumps and other places where hazardous substances have been released. PRPs include existing owners and operators, former owners and operators, transporters to the site, generators of the waste, and others. Any person found to be even partially responsible for creating the hazardous waste site can be held completely responsible for its cleanup. This law affects sites that were in existence before the legislation was passed, as well as sites that may have been legal at the time of their creation. EPA cast a wide net to encompass as many potentially responsible parties as possible under this regulation. Unknowing parties could be held responsible for thousands or even millions of dollars of clean-up costs through CERCLA.

SARA

One of the most prodigious tasks that an individual who is responsible for environmental regulations must do is gather data, file it, and in some cases, submit it to the government. These responsibilities were increased with the passage of the Superfund Amendments and Reauthorization Act (SARA) in 1986.

SARA was proposed as a more stringent response to hazardous waste releases. SARA and CERCLA address hazardous substance releases into the environment and the cleanup of inactive hazardous waste disposal sites (Noll et al, 1988, p. 5).

SARA is broken down into five Titles:

I.   Provisions Relating Primarily To Response and Liability
II. Miscellaneous Provisions
III. Emergency Planning and Community Right-To-Know
IV. Radon Gas and Indoor Air Quality Research
V.   Amendments of The Internal Revenue Code of 1986

Title I and II feature the major changes to sections of CERCLA while Title V discusses the Superfund and its revenues. Under this regulation, any company with threshold quantities of extremely hazardous substances (EHSs) must contact state and local emergency planners and the fire department. *Threshold quantities* refers to specified amounts of certain chemicals. If the company has these chemicals, it must also submit copies of its Material Safety Data Sheets (MSDSs) to the fire department and emergency planners.

If a company exceeds certain thresholds, it will have to submit chemical inventory forms to state and local fire fighters. EPA refers to these forms as Tier I and Tier II. Tier I forms are submitted by March 1 each year and Tier II forms are submitted when requested by emergency response officials.

The Occupational Safety and Health Administration (OSHA) sets requirements for the health and safety of hazardous waste workers, RCRA/TSD site workers, and emergency responders (US DOL, 1991). These requirements are found in the *Code of Federal Regulations* at 29 CFR 1910.120, Hazardous Waste Operations and Emergency Response, which is an outgrowth to Title I of the Superfund Amendment and Reauthorization Act (SARA). This regulation is also known as HAZWOPER (Hazardous Waste Operations and Emergency Response).

WORKER PROTECTION STANDARDS

Employees that work in hazardous waste operations want to be able to perform their duties safely and be assured that their health is not at risk. The law proposed standards that facilities must follow to protect the health and safety of their employees in a hazardous waste operation. Regulations were issued on these standards which include but are not limited to the following provisions:

1. Site Analysis. Requirements for a formal hazard analysis of the site and development of a site specific plan for worker protection.

2. Training Requirements. Requirements for contractors to provide initial and routine training of workers before they are permitted to engage in hazardous waste operations which could expose them to toxic substances.

3. Medical Surveillance. Requirements for a program of regular medical examination, monitoring, and surveillance of workers engaged in hazardous waste operations which could expose them to toxic substances was outlined.

4. Protective Equipment. Requirements for appropriate personal protective equipment, clothing, and respirators for individuals working in hazardous waste operations.

5. Engineering Controls. Requirements for engineering controls concerning the use of equipment and exposure of workers engaged in hazardous waste operations.

6. Maximum Exposure Limits. Requirements for maximum exposure limitations for workers engaged in hazardous waste operations, including necessary monitoring and assessment procedures.

7. Informational Programs. Requirements for a program to inform workers engaged in hazardous waste operations of the nature and degree of toxic exposure likely as a result of such hazardous waste operations.

8. Handling. Requirements for the handling, transporting, labeling, and disposing of hazardous wastes.

9. New Technology Program. Requirements for a program that introduces new equipment or technologies that will maintain worker protection.

10. Decontamination Procedures. Requirements for decontamination procedures.

11. Emergency Response. Requirements for emergency response and protection of workers engaged in hazardous waste operations.

This law also sets standards on the amount of training that employees at hazardous waste sites must have in order to protect their health and safety. Everyone who is going to be involved with the hazardous waste operations must be trained for a minimum of 40 hours of initial instruction off-site and 3 days of actual field experience under trained supervision. The employees must have additional training if the hazards that they are working with are unique or special. Supervisors must have the same training as the other employees, plus an additional 8 hours of specialized training on managing a hazardous waste operation.

Workers at Treatment Storage and Disposal (TSDs) facilities and emergency responders must have a minimum of 24 hours of initial training. Workers in all categories must complete an eight-hour refresher course annually.

Workers at points where hazardous waste is generated must also be trained in 29 CFR 1910.1200, the Hazard Communication Standard (also referred to as HAZCOM). Although MSDSs and other requirements of the standard are not required for hazardous waste generation, they must be covered relative to the chemicals that may turn into hazardous waste. HAZCOM requires that companies maintain a file of all MSDSs, label all hazardous chemicals, and train workers in the handling of hazardous chemicals.

HAZARD COMMUNICATION STANDARD (1910.1200)

Approximately 32 million U.S. workers are potentially exposed to one or more hazardous chemicals in their workplaces. There are an estimated 575,000 hazardous chemicals already found in the workplace with literally hundreds of new ones being introduced each year. The chemicals can be either health hazards, physical hazards, or in some cases, both. Simple rashes to more serious conditions such as, burns, sterility, lung damage, heart ailments, kidney damage, or cancer can be associated with exposures to hazardous chemicals (State of Kentucky, p. 1).

Protecting the employee from the harmful effects of hazardous chemical exposures is a formidable task for the employer. In order for a chemical to harm an individual's health, it must come in contact with or enter the body. The three primary routes of entry or pathways by which a chemical can enter the body are inhalation, absorption, and ingestion and injection (refer to the chapter on Industrial Hygiene). For the employer to develop and implement control measures to block these routes of entry, all pertinent information about a chemical is essential.

In 1983, the Occupational Safety and Health Administration (OSHA) enacted the "Hazard Communication Standard" that confronts the seriousness of this health and safety problem. The purpose is to ensure that the hazards from all chemicals produced or imported are evaluated and the hazard information transmitted downstream to all affected parties. The rule also guarantees the workers' right to know about the hazardous chemicals in their workplaces, therefore, it was often referred to as the right-to-know standard. After the passage of SARA, it has been typically referred to as the "Haz Com Standard."

While there is no all-inclusive list of chemicals covered by the HCS, the following lists contain chemicals considered to be hazardous:

1. Chemicals regulated by OSHA in 29 CFR Part 1910, Subpart Z, Toxic and Hazardous Substances

2. Chemicals included in the American Conference of Governmental Industrial Hygienists' (ACGIH) latest edition of *Threshold Limit Values (TLVs) for Chemical Substances and Physical Agents in the Work Environment*

3. Chemicals found to be suspected or confirmed carcinogens by the National Toxicology Program in the latest edition of the *Annual Report on Carcinogens,* or by the International Agency for Research on Cancer in the latest edition of its *IARC Monographs* (State of Kentucky, p. 1)

The following chemicals are excluded from the provisions of the standard:

1. Hazardous waste as defined by the Resource, Conservation and Recovery Act (RCRA)
2. Tobacco and tobacco products
3. Wood and wood products
4. Articles that do not release or otherwise result in exposure to a hazardous chemical under normal conditions of use
5. Food, drugs, cosmetics, or alcoholic beverages in a retail establishment
6. Food, drugs, or cosmetics intended for personal consumption by employees while in the workplace
7. Any drugs when sold in final form for direct administration to a patient (State of Kentucky, pp. 1-3)

There are four basic requirements under the HCS: the written hazard communication program, labeling, MSDSs, and employee training.

Written Program

The employer must develop, implement, and maintain at the workplace a written hazard communication program that contains at least the following information:

1. Information explaining labels and other forms of warning
2. Information on location and availability of MSDSs
3. Details of how the employee training requirements are to be satisfied
4. A list of all the hazardous chemicals known to be present in the workplace
5. The methods the employer will use to inform employees of the hazards involved in non-routine tasks and the hazards associated with chemicals contained in unlabeled pipes in their workplace areas

Organizations requiring contracted services in locations of their facility where hazardous materials are used or stored, must inform those contracted employees of all hazards present. This must be part of the written program. This program should include methods and procedures the employer will use to provide all contractors with MSDSs, information on labeling and other forms of warning, and information concerning precautions needed to protect their employees while on-site.

The employer shall make the written program available, upon request, to any employee or his designated representatives. OSHA, the Director of the National Institute for Occupational, Safety, and Health, or their designees may also request the program.

Labeling

The chemical manufacturers, distributors, or importers must ensure that each container of hazardous chemicals leaving their locale is labeled, tagged, or marked with the following information:

1. Identity of the hazardous chemical
2. Appropriate hazard warnings
3. Name and address of the manufacturer, distributor, importer, or other responsible party

However, it should be noted that other information is often required to comply with other regulations and standards beyond the scope of haz com.

The employer is not required to affix new labels, providing that the existing labels convey the required information.

The employer is required to

1. Ensure that all incoming shipments are properly labeled and the information is in English
2. Ensure that all labels on existing stock are in place, legible, and are not removed or defaced
3. Ensure that all portable containers are labeled unless it is for the immediate use of the employee that makes the transfer

MATERIAL SAFETY DATA SHEETS (MSDSs)

Chemical manufacturers shall obtain or develop an MSDS for each hazardous chemical. These MSDSs should be sent to the distributor or user prior to or with the initial chemical shipment. An MSDS must be sent with each initial shipment after each subsequent MSDS update. The employer shall maintain a current MSDS for each hazardous chemical present in the workplace. These MSDSs must be readily accessible to all affected employees on all shifts without the employees having to leave the work area.

The MSDSs may differ considerably depending on the manufacturer but must contain basically the same information. This must be in English and contain no blank spaces. Any MSDS that contains blank spaces is considered unacceptable and should be returned to the vendor, or the missing information can be requisitioned from that vendor. The complete MSDS consists of eight sections with the following information (State of Kentucky, pp. 1-3):

Section 1. Manufacturer Identity
   a.   Who makes it
   b.   The manufacturer's address
   c.   Emergency phone number and date prepared

Section 2. Hazard Ingredients
   a.   Name of the chemical and common names
   b.   Exposure limits

Section 3. Physical and Chemical Characteristics
   a.   The material's boiling point
   b.   Solubility in water
   c.   Vapor pressure
   d.   Melting point
   e.   Evaporation point
   f.   Appearance and odor of the chemical under normal conditions
   g.   Molecular weight

Section 4. Fire and Explosive Hazard Data
   a.   The chemical's flash point
   b.   Flammable limits
   c.   Upper and lower explosive limits
   d.   Protective clothing and/or respiratory equipment required to fight the fire and the type of extinguishing material to be used to fight the fire

Section 5. Reactivity Data
   a.   Stability
   b.   Conditions to avoid
   c.   Container or shelving information
   d.   Incompatibility
   e.   Hazardous decomposition or byproducts

Section 6. Health Hazards
   a.   Specific hazard (carcinogen, neurotoxin, reproductive toxin, etc.)
   b.   Routes of entry
   c.   Medical and first aid treatment
   d.   Signs and symptoms of exposure and whether the effects are chronic or acute

Section 7. Precaution for Safe Handling and Use
   a.   How to store the chemical
   b.   What to do in the event of a spill
   c.   How to dispose of the chemical
   d.   Environmental protection

Section 8. Control Measures
   a.   Engineering controls
   b.   Personal protective equipment
   c.   Care and disposal of contaminated clothing and equipment
   d.   Any special work or hygiene practices that should be followed

Although OSHA details the information that is required to be disclosed on an MSDS, it does not prescribe a format or organization. Therefore, MSDSs appear in a number of different formats. To address this problem, the American National Standards Institute (ANSI) and the Chemical Manufacturers Association (CMA) developed Z400.1 which has now been proposed to OSHA. It requires the following sections:

1. Chemical Product and Company Information
2. Composition/Information on Ingredients
3. Hazards Identification including
   - Emergency Overview
   - Potential Health Effects
   - Chronic Effects/Carcinogenicity
4. First Aid Measures
5. Fire Fighting Measures
6. Accidental Release Measures
7. Handling and Storage
8. Exposure Controls/Personal Protection
9. Physical and Chemical Properties
10. Stability and Reactivity
11. Toxicological Information
12. Ecological Information
13. Disposal Considerations
14. Transport Information
15. Regulatory Information
16. Other Information (Chandler,1995, pp. 1431-1433)

Employee training must be conducted during the initial assignment to a work area containing one or more hazardous chemicals and each time a new hazard is introduced into the workplace. The employee training shall include the following:

1. The details of the written hazard communication program
2. The location and availability of MSDSs
3. An explanation of the labeling system and other forms of warning such as, alarms
4. The physical hazards and health hazards of the chemicals present in the employee's work areas
5. The methods and observations that may be used to detect the release of a hazardous chemical in the employee's work area

6. The measures the employees can take to protect herself from chemical hazards
7. How the employee can obtain and use the appropriate information

CONTINGENCY PLANS

If hazardous waste is stored, steps must be taken to avoid fire, explosion, spill, and release. It is necessary to have an emergency communication system, appropriate fire equipment, and plans with local police and emergency response teams. (Kaufman, 1990).

Contingency plans must be drawn up that include emergency procedures, emergency phone numbers, name of the emergency coordinator, and containment measures available. Training records for all emergency responders must be maintained (US DOL, 1991).

HAZARDOUS WASTE DISPOSAL

Hazardous waste may not be disposed of on-site unless a disposal permit has been obtained. Obtaining a permit to store, treat, or dispose of hazardous waste on-site (40 CFR Part 270) can be costly and time consuming.

The first step in the proper disposal of hazardous waste off-site is to select a licensed hazardous waste hauler and disposal facility. Companies must ship waste with haulers and to facilities with EPA ID numbers. Either the waste generator or the hauler must package and label wastes for shipping and prepare a Hazardous Waste Manifest (Kaufman, 1990).

When hazardous waste is shipped off-site, it is still the responsibility of the generator. Incineration of the waste, by a licensed treatment facility, is usually the selection of choice due to the fact that the waste is destroyed and no future liability can be incurred. Land disposal requires land ban exemption forms to be filed.

UNIFORM HAZARDOUS WASTE MANIFEST AND DOT REGULATIONS

The Department of Transportation (DOT) regulations of 49 CFR Part 172 require hazardous waste to be properly packaged in containers that are acceptable for transportation on public roads. Proper labeling of containers includes the identification of the waste, the words "Hazardous Waste," and the DOT waste codes and placards. The Department of Transportation also requires a Uniform Hazardous Waste Manifest. This is a six-part shipping document that must accompany all hazardous waste shipments. It is designed so that hazardous waste can be tracked from the generator to the final destination or disposal or "cradle-to-grave." The waste in the shipment must be the same as the waste listed on the manifest.

ADDITIONAL REPORTING REQUIREMENTS

Hazardous Waste Annual Report Forms must be submitted annually to the USEPA or state EPA (Chapter 31, 1992). The annual reporting form includes information on all the hazardous waste generated and shipped during the report year. All facilities that treat, dispose, or recycle hazardous waste on-site must complete the form. Waste minimization activities must be identified on the form.

EPA WEARS MANY HATS

The EPA plays several roles. Acting like a miniature Congress, it develops and issues its own regulations. These regulations can be, and usually are, much more comprehensive than the original legislation. The EPA is divided into three different divisions: Administration Division, Division of Compliance, and Division of Enforcement.

The Administration Division provides technical assistance to hazardous waste generators and TSD operators. This division is especially helpful with recycling and waste minimization efforts (Freeman, 1990).

The EPA also acts as a policeman conducting inspections and investigating incidents. Inspections of hazardous waste generation facilities and TSDs are conducted by the Division of Compliance. The compliance officer will inspect the facility and fill out an inspection report. During the inspection, the compliance officer will do a walk-through tour of the facility, take pictures of any violations, and review the necessary paperwork. The compliance officer will have a closing interview with the operator of the facility and point out any violations. A Notice of Violation

(NOV) will be sent by certified mail. Any violations must be corrected within a time period specified on the NOV. The compliance officer will conduct a follow-up inspection to make sure all the violations are corrected.

At a certain level, the EPA acts as judge and jury. The Division of Enforcement is contacted when a violation is not corrected within the given time period or when penalties for a violation are assessed. Enforcement actions can include civil penalties or criminal charges, which can lead to fines and/or imprisonment. If an EPA inspection does result in a citation, the facility may choose to contest the citation through an appeals process within EPA instead of paying the fine. The facility has 30 days after receiving the NOV to contest the citation.

The vast majority of federal environmental issues are handled at the EPA level. There are, however, some critical cases that do wind up in the federal courts. Many environmental statutes contain provisions for criminal penalties. Violations of the criminal provisions are prosecuted in the courts rather than through administrative actions. Under certain, rather narrow circumstances, companies may appeal adverse EPA decisions on regulation violation to the courts.

CONCLUSION

Compliance with regulations that govern the handling, storage, transportation, and disposal of hazardous materials is a critical component of any safety practitioner's job. The mishandling of hazardous materials can be costly in a number of ways. Monetary losses and human suffering can result from poorly planned actions on the part of untrained personnel. Violators can find themselves at the mercy of EPA, OSHA, and the courts. The proactive safety practitioner should become familiar with existing regulations at both the federal and state levels and make every effort to stay apprised of changes in the law as they affect the workforce.

Questions

1. What hazardous chemicals are located in your area? Contact your local fire department and tell them you wish to learn about hazardous chemicals in your area. Ask them if copies of MSDSs are available and where they can be found.

2. What are the responsibilities of environmental managers? Find someone who has environmental responsibilities for his company and ask what his major responsibilities are. Ask specifically about filing MSDSs with the fire department and emergency planning officials. Also ask about what training is required for workers to respond to spills of any hazardous chemicals on the property. Ask this person which CFRs are available on-site and if you can take a look at them. Some companies now purchase these on CD-ROM, so it's hard to get a feel for the scope of the regulations.

3. Why do you think many companies complain about the difficult legal framework and red tape they face in trying to comply with all applicable regulations? Take a look at some of the individual CFRs in your library.

References

Bowman, V. 1988. *Checklist for Environmental Compliance.* 4th ed. pp. 24-29. Newton, MA: Cahners.

Chandler, R. L., Iannaccone, M., and Toki, A. Eds.. 1995. *Best's Safety Directory: Industrial Safety, Hygiene, Society.* 1995 ed., Vol. 2. Oldwick, NJ: A. M. Best Company.

Freeman, H. 1990. *Hazardous waste minimization.* 1st ed. pp. 3-9. New York, NY: McGraw-Hill.

Kaufman, J. Ed.. 1990. *Waste Disposal in Academic Institutions.* 2nd ed.. Chelsea, MI: Lewis.

Noll, G. G., Hildebrand, M. S. and Yvorra, J. G. 1995. *Hazardous Material: Managing the Incident.* Stillwater, OK: Fire Protection Publications.

State of Kentucky. 1992. Waste Management-identification and listing of hazardous wastes: Title 401. *Kentucky Waste Management Regulations.* Frankfort, KY.

U.S. Department of Labor. 1991. *Hazardous Waste Operations and Emergency Response*: 29 CFR. 1910.120. Washington, DC: U.S. Government Printing Office.

U.S. Environmental Protection Agency. 1991. *Protection of the Environment*: Title 40 CFR. Washington, DC: U.S. Government Printing Office.

Chapter 14

Resources for Safety and Health

The following is a noncomprehensive list of resources available to the safety professional seeking assistance in selected areas of occupational safety and health. There are numerous resources available that have not been included in this list because of space limitations. Inclusion in the following list is not intended to be an endorsement of individuals, organizations, products, or services. Resources are listed under the broad topic areas that comprise the area of occupational safety and health.

PROFESSIONAL ORGANIZATIONS AND AGENCIES

ABIH (American Board of Industrial Hygiene)
4600 West Saginaw; Suite 101
Lansing, Michigan 48917
(517) 321-2638

The board is responsible for the certification of industrial hygienists. It oversees the examination process and credentialing of industrial hygiene practitioners.

ACGIH (American Conference of Governmental Industrial Hygienists)
Building D-7
6500 Glenway Avenue
Cincinnati, Ohio 45211
(513)661-7881

Professional society of persons employed by official governmental units responsible for full-time programs in industrial hygiene. Devoted to the development of administrative and technical aspects of worker health protection.

AIHA (American Industrial Hygiene Association)
P.O. Box 8390
475 White Pond Drive
Akron, Ohio 44311
(216) 873-3300

Professional society for industrial hygienists. Promotes the study and control of environmental factors affecting the health and well-being of industrial workers.

ANSI (American National Standards Institute)
11 West 42 Street
New York, New York 10038
(212) 354-3300

Serves as a clearing house for nationally coordinated voluntary safety, engineering, and industrial standards.

ASME (American Society of Mechanical Engineers)
345 East 47th Street
New York, New York 10017
(212) 705-7722

Conducts research and develops boiler, pressure vessel, and power test codes. Sponsors American National Standards Institute in developing safety codes and standards for equipment.

ASSE (American Society for Safety Engineers)
1800 East Oakton Street
Des Plaines, Illinois 60016
(708) 692-4121

Professional society of safety engineers, safety directors, and others concerned with accident prevention and safety programs.

ASTM (American Society for Testing and Materials)
655 15th Street NW
Washington, D.C. 20005
(202) 639-4025

Establishes voluntary consensus standards for materials, products, systems, and services.

AWS (American Welding Society)
P.O. Box 351040
550 LeJeune Road, NW
Miami, Florida 33135
(305) 443-9353

Professional engineering society in the field of welding. Sponsors seminars and conferences on welding.

BCPE (Board of Certification in Professional Ergonomics)
P.O. Box 2811
Bellingham, WA 98227-2811

The BCPE was established to provide a formal organization and procedures for examining and certifying qualified practitioners of ergonomics.

BCSP (Board of Certified Safety Professionals)
208 Burwash Ave.
Savoy, IL 61874
(217) 359-9263

The board is responsible for the certification of safety professionals. It oversees the examination process and credentialing of practitioners in the safety field.

BLS (Bureau of Labor Statistics)
U.S. Dept. of Labor
Occupational Safety and Health Statistics
441 G Street, NW
Washington, D.C. 20212
(202) 523-1382

This federal agency gathers workplace-related statistical information.

CDC (Center for Disease Control)
U.S. Dept. of Health and Human Services
1600 Clifton Avenue, N.E.
Atlanta, Georgia 30333
(404) 329-3311

Surveys national disease trends and epidemics and environmental health problems. Promotes national health education programs. Administers block grants to states for preventive medicine and health services programs.

CGA (Compressed Gas Association)
1235 Jefferson Davis Highway
Arlington, Virginia 22202
(703) 979-0900

Submits recommendations to appropriate government agencies to improve safety standards and methods of handling, transporting, and storing gases. Acts as an advisor to regulatory authorities and other agencies concerned with safe handling of compressed gases.

EPA (Environmental Protection Agency)
410 M Street, S.W.
Washington, D.C. 20460
(202) 382-4361

Administers federal environmental policies, research, and regulations. Provides information on many environmental subjects including water pollution, hazardous and solid waste disposal, air and noise pollution, pesticides, and radiation.

Superintendent of Documents
U.S. Government Printing Office (GPO)
732 N. Capitol Street, NW
Washington, D.C. 20402
(202) 235-1452

Prints, distributes, and sells selected publications of the U.S. Congress, government agencies, and executive departments.

MSHA (Mine Safety and Health Administration)
U.S. Department of Labor
4015 Wilson Blvd.
Arlington, Virginia 22203
(703) 235-1452

Administers and enforces the health and safety provisions of the Federal Mine Safety and Health Act of 1977. Its training facility is located in Beckley, West Virginia.

NAC (National Audio Visual Center)
National Archives and Records Administration
Customer Services Section CL
8700 Edgewood Drive
Capitol Heights, Maryland 20743-3701
(301) 763-1896

Serves as the central source for all federally produced audiovisual materials and makes them available to the public through information and distribution services.

National Institute of Standards and Technology
U.S. Dept. of Commerce
National Engineering Laboratory
Route I-270 and Quince Orchard Road
Gaithersburg, Maryland 20899
(310) 921-3434

Develops engineering measurements, data, and test methods. Produces the technical base for proposed engineering standards and code changes. Generates new engineering practices. Aids international competitiveness of small- and medium-size companies and consortia through technology development and transfer programs.

NFPA (National Fire Protection Association)
1 Batterymarch Park
Quincy, Massachusetts 02269
(800) 344-3555

Develops, publishes, and disseminates standards prepared by approximately 175 technical committees, intended to minimize the possibility and effects of fire and explosion.

NIH (National Institutes of Health)
U.S. Department of Health and Human Services
9000 Rockville Pike
Bethesda, Maryland 20205
(310) 496-5787

Supports and conducts biomedical research into the causes and prevention of diseases and furnishes information to health professionals and the public.

NIOSH (National Institute for Occupational Safety and Health)
U.S. Dept. of Health and Human Services
Publications Dissemination
4676 Columbia Parkway
Cincinnati, Ohio 45226
(513) 533-8287

Part of the Center for Disease Control. Supports and conducts research on occupational safety and health issues. Provides technical assistance and training. Develops recommendations for OSHA.

NSC (National Safety Council)
444 North Michigan Avenue
Chicago, Illinois 60611
(312) 527-4800

A voluntary, nongovernmental organization that promotes accident reduction by providing a forum for the exchange of safety and health ideas, techniques, experiences, and accident-prevention methods.

NTIS (National Technical Information Services)
U.S. Dept. of Commerce
5285 Port Royal Road
Springfield, Virginia 22161
(703) 487-4636

A distribution center that sells to the public government-funded research and development reports prepared by the federal agencies, their contractors, or grantees. Offers microfiche and computerized bibliography search services.

OSHA (Occupational Safety and Health Administration - National Office)
U.S. Dept. of Labor
200 Constitution Avenue, NW
Washington, D.C. 20210
(202) 523-8151

Sets policy, develops programs, and implements the Occupational Safety and Health Act of 1970.

OSHRC (Occupational Safety and Health Review Commission)
1825 K Street, NW
Washington, D.C. 20006
(202) 643-7943

Independent executive agency that adjudicates disputes between private employers and OSHA, arising from citations of occupational safety and health standards.

Women's Occupational Health Resource Center
117 St. Johns Place
Brooklyn, New York 11217
(718) 230-8822

Acts as a clearing house for women's occupational health and safety issues.

ERGONOMICS

Anthropometrics

Publication Resources

Book: *Bodyspace: Anthropometry, Ergonomics, and Design* / S. T. Pheasant, London: Taylor & Francis (1986)

Behavioral and Psychological Aspects of Ergonomics

Associations and Organizations

Behavioral Science Technology, Inc.
323 Least Matilija Street, Suite 215
Ojai, CA 93023
(800) 548-5781
Provides training in behaviorally based safety programs.

Buffalo Organization For Social and Technological Innovation (BOSTI)
1479 Hertel Ave.
Buffalo, NY 14216
(716) 837 - 7120
Design, research, and consulting associated with the behavior, performance, and satisfaction of people.

Project On Technology, Work and Character (PTWC)
1700 K St. NW, Ste. 306
Washington, DC 20006
(202) 296 - 4300
Identifies factors making work satisfying and productive.

Biomechanics

Associations and Organizations

MTM Association For Standards And Research (MTM)
1411 Peterson Ave.
Park Ridge, IL 60068
(708) 823 - 7120

Electronic Resources

Internet info@webcrawler.com"Biomechanics"
U.S. Occupational Health, Inc.
205 W. Randolph St., Ste. 720
Chicago, IL 60606
(800) 548 - 5909
Software to predict the static strength needed to perform a variety of material-handling actions. Used to predict which employees can perform a task safely.

Publication Resources

Journal: *Journal Of Human Movement Studies*
Edinburgh, Scotland: Teviot Scientific Publications
031 - 332 - 8764
An academic/scholarly publication.

Book: *The Biomechanical Basis of Ergonomics: Anatomy Applied to the Design of Work Situations* / E. R. Tichauer, New York: Wiley (1978)
A book on human engineering and biomechanics.

Book: *Engineering Physiology: Bases of Human Factors/Ergonomics* / K.H.E. Kroemer, H. J. Kroemer-Elbert, New York: Van Nostrand Reinhold (1990)
A book that presents physiological information related to human factors and ergonomics.

Ergonomics Engineering--Workstation Layout and Design

Associations and Organizations

Operations Management Education And Research Foundation (OMER)
PO Box 661
Rockford, MI 49341
(616) 732 - 5543
An organization dedicated to the open exchange of ideas, information, and assistance regarding the work environment.

Consultants

Biomechanics Corporation of America
1800 Walt Whitman Rd.
Melville, New York 11747
516 752-3550
800 969-5123
516 752-3506 (fax)

Ergonomiks
P.O. Box 458 3738 E. Commonwealth Pl., #14
Chandler, AZ 85244
(602) 821 - 0721

Human Dynamics, Inc.
127 W. Sixth Ave., Ste. C
Lancaster, OH 43130
(800) 767 - 6211

Humantech Inc.
173-T Parkland Plaza, Suite D
Ann Arbor, Michigan 48103
800 959-0650
313 663-7747 (fax)

The Joyce Institute (Training and Consulting)
1313 Plaza 600 Bldg.
Seattle, WA 98101
800 645-6045

USE Ergonomic Services Group
4401 Dayton-Xenia Rd.
Dayton, Ohio 45432-1894
513 426-6900, ext. 180

National Safety Council
1121 Spring Lake Drive
Itasca, IL 60143
(800) 621 - 7619

Manufacturers and Suppliers

Broner Glove and Safety
1750 Hamon Rd.
Auburn Hills, MI 48326
(800) 521 - 1318
Custom-made, ergonomically designed work stations for assembly line workers.

Safety Products, Inc.
P.O. Box 658
Eaton Park, FL 33840
(813) 665 - 3601
Custom-made, ergonomically designed work stations for assembly line workers.

Publication Resources

Book: *The Ergonomics of Workspaces and Machines: A Design Manual/* T. S. Clark and E. N. Corlett, London: Taylor & Francis (1984)

Journal: *Ergonomics*
London: Taylor & Francis
An international journal devoted to the study of human factors in relation to working environment and equipment design.

Ergonomics Engineering--Hand and Power Tools

Manufacturers and Suppliers

U.S. Occupational Health, Inc.
205 W. Randolph St., Ste. 720
Chicago, IL 60606
(800) 548 - 5909
An ergonomic, task-tool-interaction measurement system used to quantify the effects of stress on the human hand and wrist.

Orr Safety Corporation
P.O. Box 16326, 2360 Millers Ln.
Louisville, KY 40256
(800) 669 - 1677
Supplier of high-friction thermoplastic material applied to tool handles and then molded to the shape of the user's hand to provide an ergonomic grip.

Ergonomics Management

Associations and Organizations

International Association For Time Use Research
Saint Mary's Univ. Halifax, NS
Canada B3H 3C3
(902) 496 - 8728
Facilitates communication and exchange of information in time and motion studies, ergonomics, and industrial efficiency.

International MTM Directorate (IMD)
PO Box 710 S-645 59
Strangnas Sweden
152 21520
Works to guarantee uniform standards of MTM practice by coordinating training and examination development and through publication of standards.

Consultants

Crawford & Company
5620 Glenridge Dr.
Atlanta, GA 30342
(404) 256 - 0630
Human-factor engineers who design or alter existing work place systems.

National Safety Council
1121 Spring Lake Drive
Itasca, IL 60143
(800) 621 - 7619
Consultants and designers of workplace systems.

Sweda Risk Services, Inc.
881 Third St., Ste. B-8, P.O. Box 227
Whitehall, PA 18052
(215) 266 - 8901
Coordinates medical treatment and rehabilitation for debilitating injuries sustained on the job. Prompt and appropriate medical treatment is given while workers' compensation costs are reduced.

Electronic Resources

UES Ergonomic Services Group
4401 Dayton-Xenia Rd.
Dayton OH 45432
(513) 426 - 6900
PC software to help employers analyze task, tool, and employee interaction to ergonomically analyze job.

Manufacturers and Suppliers

BNA Communications Inc.
9439 Key West Ave.
Rockville, MD 20850
(800) 217 - 2338
Audiovisual programs that train workers how to prevent repetitive-motion problems.

The Marcom Group Ltd.
#4 Denny Rd.
Wilmington, DE 19809
(800) 654 - 2448
Audiovisual, computer programs, and literature detail ways the workplace might be altered to alleviate fatigue-related pain for workers.

Publication Resources

A.M. Best Co., *Best's Safety Directory*
Ambest Rd.
Oldwick, NJ 08858
(908) 439 - 2200
Reference manual for safety products and services.

Conney Safety Products
3202 Latham Dr.
Madison,WI 53713
(608) 271 - 3300
An ergonomics manual featuring an index of organizations, associations, databases, research centers, consultants, and periodicals.

Book: *Men At Work; Applications of Ergonomics to Performance and Design*
Roy J. Shepard
Springfield, IL: Thomas (1974)
Subjects include human engineering and the physiological aspects of work.

Ergonomics--General

Associations and Organizations

Human Factors and Ergonomics Society (HFES)
PO Box 1369
Santa Monica, CA 90406-1369
(310) 394 - 1811
Professional association concerned with the use of human factors and ergonomics in the development of systems and devices.

Center For Ergonomics Research
Miami University, Dept. of Psychology
104 Benton Hall
Miami, OH 45056
(513) 529 - 2414
Conducts research and educational programs on ergonomics.

International Ergonomics Association (IEA)
Central Advisory Group on Ergonomics, Personnel Affairs Dept.
Netherlands Railways NL - 3511
Utrect Netherlands
30 354455
Federated societies in 27 countries, interested in the scientific study of human work and its environment.

American Society of Safety Engineers
1800 E. Oakton St.
Des Plaines, IL 60018
(708) 692 - 4121

Commerce Clearing House Inc.
Employment Safety & Health Guide
4025 W. Peterson Ave.
Chicago, IL 60646
(312) 583 - 8500

Electronic Resources

Internet info@webcrawler.comErgonomics
The following databases are available at ECU: Proquest, Infotrac, General Science Index, Applied Science and Technology Index, Predicasts F & S Index, Psyclit, Citation Index, Sociological Abstract.

Publication Resources

Book: *Ergonomics Sourcebook: A Guide To Human Factors Information* / Kimberlie H. Pelsma, editor
Lawrence, KS : Report Store (1987)

Book: *Cumulative Trauma Disorders: Current Issues and Ergonomic Solutions: A Systems Approach* Kathryn G. Parker and Harold R. Imbus
Boca Raton: Lewis Publishers (1992)

Newsletter: *Ergotalk*
Raleigh, NC: North Carolina Resource Center
(919) 515 - 2052

State and Federal Agencies

Occupational Safety and Health Administration
U.S. Department of Labor
200 Constitution Ave.
Washington, DC 20210
(202) 523 - 7075

National Institute for Occupational Safety & Health
1600 Clifton Rd., N.E.
Atlanta, GA 30333
(800) 356 - 4674

North Carolina Ergonomics Resource Center
703 Tucker St.
Raleigh, NC 27603
(919) 515 - 2052

INDUSTRIAL HYGIENE

Air Pollution and Hazardous Waste

Associations and Organizations

Center for Hazardous Material Research
320 William Pitt Way
Pittsburgh, PA 15238
(412) 826-5320
This multinational organization seeks to develop practical solutions to problems facing hazardous and solid-waste management. Working in conjunction with industry and government agencies, CHMR conducts programs on the use and disposal of hazardous waste

Citizens Clearinghouse For Hazardous Waste
PO Box 6806
Falls Church, VA 22040
(703) 237-2249

This organization is concerned with the physical effects of toxic chemicals that come into contact with people. It also provides information and guidance in dealing with hazardous problems.

Clean Sites, Inc.
1199 N. Fairfax St.
Alexandria, VA 22314
(703) 683-8522
Organized to improve the process of cleaning up waste. CSI also provides alternate dispute resolutions, public policy analysis, and technical project management.

Consultants

Applied Health Physics
2986 Industrial Blvd.
Bethel Park, Pa 15102
(412) 835-9555
These services provide assistance in the development of hazardous waste disposal. The personnel are also capable in the handling, packaging, etc., of dangerous materials.

AXIA Services, Inc.
151 Farmington Ave.
W101, Hartford, Ct. 06156
(203) 683-3624
Provide assistance in the development of hazardous waste disposal. Also provide training.

Brown and Root Environment
Foster Plaza 7
661 Anderson Dr.
Pittsburgh, PA. 15220.
(412) 921-7090
These services are used to provide assistance in the development of hazardous waste disposal. They have trained personnel capable in the in handling, packaging, etc., of dangerous materials.

Chemical Safety Associates, Inc.
9163 Chesapeake Drive
San Diego, Ca. 92123
(619) 565-0302
These services provide assistance in the development of hazardous waste disposal. They have trained personnel capable in handling, packaging, etc., of dangerous materials.

Electronic Resources

Internet info@ webcrawler.comair pollution/hazardous waste

Manufacturers and Suppliers

Sensation Incorporated
16333 Bay Vista Drive
Clearwater, FL 34620
(800) 530-3602
Manufactures wide range of air monitoring equipment.

S.C. Incorporated
863 Valley View Road
Eighty Four, PA 15330
(800) 752-8472
Manufactures wide range of air supplies and equipment.

American Health and Safety Inc.
6250 Nesbitt Rd.
Madison, WI. 53719
(800) 522-7554
Manufactures disposable mask that filters out unpleasant or nontoxic odors. Mask does not restrict breathing or hold in heat.

New Pig
One Pork Avenue
Tiptoe, PA 16684-0304
(800) 643-6465
Manufactures hazardous-material spill clean-up equipment.

Health Physics and Radiation

Associations and Organizations

Association For Radiation Research
Mount Vernon Hospital
PO Box 100
Northwood
Middlesex, Greater London HA6 2JR England 1923 828611
Scientific, medical, etc. professionals who are interested in ionizing radiation. Organizes research and grants money to students who wish to attend the AAR conference.

Center For Atomic Radiation Studies
PO Box 381036
Cambridge, MA 02238
(508) 263-2065
Conducts research on the effects of ionizing radiation on people. Works with people who may have been exposed due to atomic release.

International Commission On Radiation Units And Measurements.
7910 Woodmont Ave.,
Bethesda, MD 20814
(301) 657-2652
Develops recommendations on the dosage needed and allowed during radioactive procedures on or near the human body.

Consultants

Applied Health Physics
2986 Industrial Blvd.
Bethel Park PA, 15102
(412) 835-9555
Specialists who clean and inspect radiation protection equipment. They can detect causes of failure in machines.

Brown and Root Environmental
661 Anderson Dr.
Pittsburgh, PA 15220
(412) 921-7090
Available for all types of decontamination procedures. Training can be
provided if requested.

Toxicology International
PO Box 75477
Washington, DC. 20013
These professionals set up monitoring programs with companies that
have high levels of radiation.

Manufactures/Suppliers

RTCA-Radon Testing Corp.
PO Box 258
Irvington, NY 10533
This device ensures that a radon monitor is accurate. This way you do
not have to send away for verification.

F.W. Bell Inc
6120 Hanging Moss Rd.
Orlando, Fl. 32807
(407) 978-6900
This device measures ELF radiation and is good for in-plant or remote
locations.

Stress

Associations and Organizations

International Society For Traumatic Stress Studies
60 Revere Dr.
Northbrook, IL 60062
(708) 480-9080
Professionals who treat individuals who suffer from stress, and conduct
research.

International Stress Management Association
10455 Pomerado Rd.
San Diego, CA 92131
(619) 693-4698
Conducts research on tension control and to incorporate relaxation in everyday life.

Toxicology

Associations and Organizations

American Board of Toxicology
PO Box 30054
Raleigh, NC 27622
(919) 782-0036
This association certifies toxicologists and administers annual certification exams.

Genetic Toxicology Association
1111 General Sullivan Rd.
Washington Crossing, PA 18977
Members are toxicologists who seek to keep informed about recent developments in genetic toxicology.

Society of Toxicology
1767 Business Center Dr.
Reston, VA 22090-5332
(703) 438-3115
Members have published investigations into toxicology and must have a continuing interest in the field.

Consultants

Chemical Safety Associates, Inc.
9163 Chesapeake Drive,
San Diego, CA 92123
(619) 565-0302

Assists public safety directors in determining the extent of the hazard. These specialists will help determine whether any of the substances are toxic.

Environmental Quality Consultants
1300 Memory Lane
Columbus, OH, 43209
(614) 252-3621
Specialists who can help safety directors in determining the risk of a hazard and suggest ways in bringing problems under control.

ERM Group, The
855 Springdale Drive
Exton, PA, 19341
(800) 544-3117
Specialists who can determine the extent of hazards in the workplace. If substances are determined to be toxic, then they can implement ways to bring the problem under control.

Manufacturers and Suppliers

Broner Glove and Safety
1750 Harmon Rd.
Auburn Hills, MI 48326
(800) 521-1318
These reference manual sheets are useful in providing background information on toxic substances and to compiling documentation relating to toxic information.

Lab Safety Supply, Inc.
401 S. Wright Rd.
Janesville, WI, 53546
(800) 356-0783
These reference manual sheets are useful in providing background information on toxic substances and to compile documentation relating to toxic information.

Noise Control

Associations and Organizations

Noise Control Association
680 Rainer Ln.
Port Ludlow, WA 98365-9775
Consists of companies that distribute noise-control products. They also offer seminars on noise control.

Institute of Noise Control Engineering.
PO Box 3206
Poughkeepsie, NY 12603
(914) 462-4006
Members consist of noise-control professionals. Develops engineering solutions to environmental noise problems.

National Organization To Insure A Sound-Controlled Environment
1225 Eye St. NW
Washington, DC 20005
(202) 682-3901
Seeks to eliminate jet noise from the environment by confronting legislature to ensure that rules are being followed.

Manufacturers and Suppliers

Bruel and Kjaer Instruments
721 N. Eckhoff St.
Orange, CA 92668
(714) 978-8066
This company manufactures a device that measures the various energies of sounds on several levels

Ametek Mansfield and Green Division
8600 Somerset Dr.
Largo, FL 34643
(813) 536-7831
This device has a calibrator as well as a sound-level meter for overall noise measurements.

CEL Instruments Ltd.
1 Westchester Dr.
Milford, NH 03055
(800) 366-2966
Manufactures noise dosimeters which can be worn by personnel to obtain individual exposure samples.

Quest Technologies
510 South Worthington
Oconomowoc, WI 53066
(800) 245-0779
Manufactures noise-monitoring equipment including octave band analyzers, dosimeters, and sound-level meters.

North Safety Products
Siebe North, Inc.
2000 Plainfield Pike
Cranston, RI 02921
Manufactures a wide range of personal protective equipment including respiratory and hearing protection.

Elvex Corporation
7 Trowbridge Drive
Bethel, CT 06801-0850
(203) 743-2488
Manufactures a wide range of personal protective equipment including vision and hearing protection.

Consultants
Acoustical System Inc
332 James Bohaman Memorial Dr.
PO Box 146 B
Vendalia, OH 45377
This firm provides total audiometric testing programs. May supplement existing programs with services and equipment.

OCCUPATIONAL HEALTH AND SAFETY

Federal and State Regulatory Compliance

Agencies with an asterisk (*) are those locations which have a state safety and health plan under section 18(b) of the OSHAct. This agency may be contacted directly for specific information regarding regulations in the state.

Federal Regional Offices:

Region I
(CT*, MA, ME, NH, RI, VT*)
133 Portland Street
1st Floor
Boston, MA 02114
(617) 565-7164

Region II
(NJ, NY*, PR*, VI*)
201 Varick Street
Room 670
New York, NY 10014
(212) 337-2378

Region III
(DC, DE, MD*, PA, VA*, WV)
Gateway Bldg, Suite 2100
3535 Market Street
Philadelphia, PA 19104
(215) 596-1201

Region IV
(AL, FL, GA, KY*, MS, NC*, SC*, TN*)
Peachtree Street N.E.
Suite 587
Atlanta, GA 30367
(404) 347-3573

Region V
(IL, IN*, MI*, MN*, OH, WI)
230 South Dearborn
Room 3244
Chicago, IL 60604
(312) 353-2220

Region VI
(AR, LA, NM*, OK, TX)
525 Griffin Street
Room 602
Dallas, TX 75202
(214) 767-4731

Region VII
(IA*, KS, MO, NE)
911 Walnut Street
Kansas City, MO 64106
(816) 426-5861

Region VIII
(CO, MT, ND, SD, UT*)
Federal Bldg., Room 1576
1961 Stout Street
Denver, CO 80294
(303) 844-3061

Region IX
(AZ*, CA*, HI*, NV*)
71 Stevenson Street
Room 415
San Francisco, CA 94105
(415) 744-6670

Region X
(AK*, ID, OR*, WA*)
1111 Third Avenue
Suite 715
Seattle, WA 98174
(206) 553-5930

State Compliance Offices

Alabama Department of Labor
600 Administrative Building
Montgomery, Alabama 36130
(202) 261-3460

*Alaska Department of Labor
Research and Analysis Section
P.O. Box 1149
Juneau, Alaska 99802
(907) 465-4520

Territory of American Samoa
Department of Manpower Resources
Government of American Samoa
Pago Pago, American Samoa 96799
633-5849

*Industrial Commission of Arizona
1601 W. Jefferson St.
P.O. Box 19070
Phoenix, Arizona 85005
(602) 255-3739

Workers' Compensation Commission (OSH)
Arkansas Department of Labor
OSH Research and Statistics
Rm. 502, 1022 High St.
Little Rock, Arkansas 72202
(501) 371-2770

*California Department of Industrial Relations
Labor Statistic and Research
P.O. Box 603
San Francisco, California 94901
(415) 557-1466

*Colorado Department of Labor and Employment
Division of Labor
1313 Sherman St. Room 323
Denver, Colorado 80203
(303) 866-3748

*Connecticut Department of Labor
200 Folly Brook Blvd.
Wethersfield, Connecticut 06109
(203) 566-4380

Delaware Department of Labor
Division of Industrial Affairs
820 N. French St. 6th Floor
Wilmington, Delaware 19801

Florida Department of Labor and Employment Security
Division of Workers' Compensation
2551 Executive Center Circle West, Room 204
Tallahassee, Florida 32301-5014
(904) 488-3044

Guam Department of Labor
Bureau of Labor Statistics
P.O. Box 23548
Guam Main Facility
Agana, Guam 96921
477-9241

*State of Hawaii
Department of Labor and Industrial Relations
P.O. Box 3680
Honolulu, Hawaii 96811
(808) 548-7638

*Indiana Department of Labor
State Office Building Room 1013
100 N. Senate Avenue
Indianapolis, Indiana 46204
(317) 232-2693

*Iowa Bureau of Labor
307 East 7th Street
Des Moines, Iowa 50319
(515) 281-5151

Kansas Department of Health and Environment
Division of Policy and Planning
Occupational Safety and Health
Topeka, Kansas 66620
(913) 862-9360 ext 280

*Kentucky Labor Cabinet
Occupational Safety and Health Program
U.S. 127 South Building
Frankfort, Kentucky 40601
(502) 564-3100

Louisiana Department of Labor
Office of employment security
1001 North 23rd and Fuqua
Baton Rouge, Louisiana 70804
(504) 342-3126

Maine Department of Labor
Bureau of Labor Standards
State Office Building
Augusta, Maine 04330
(207) 289-3331

*Maryland Department of Licensing and Regulation
Division of Labor and Industry
501 St. Paul Pl.
Baltimore, Maryland 21202
(310) 659-4202

Massachusetts Department of Labor and Industries
Division of Industrial Safety
100 Cambridge Street
Boston, Massachusetts 02202
(617)727-3593

*Michigan Department of Labor
7150 Harris Drive, Secondary Complex
P.O. Box 30015
Lansing, Michigan 48909
(517) 322-1848

*Minnesota Department of Labor and Industry IMSD
444 Lafayette Road, 5TH floor
St. Paul, Minnesota 55101
(612) 296-4893

Mississippi State Department of Health
Office of Public Health Statistics
P.O. Box 1700
Jackson, Mississippi 39215-1700
(601) 354-7233

Missouri Department of Labor and Industrial Relations
Division of Workers' Compensation
P.O. Box 58
Jefferson City, Missouri 65102
(314) 751-4231

Montana Department of Labor and Industry
Workers' Compensation Division
5 South Last Chance Gulch
Helena, Montana 59601
(406) 444-6515

Nebraska Worker's Compensation Court
State Capitol, 12th Floor
Lincoln, Nebraska 68509-4967
(402) 471-3547

*Nevada Department of Industrial Relations
Division of Occupational Safety and Health
1370 South Curry St
Carson City, Nevada 89710
(702) 885-5240

New Jersey Department of Labor and Industry
Division of Planning and Research
C N 056
Trenton, New Jersey 08625
(609) 292-8997

New Mexico Health and Environment Department
P.O. Box 968 Crown Building
Sante Fe, New Mexico 87504-0968
(505) 827-5271 ext 230

New York Department of Labor
Division of Research and Statistics
2 World Trade Center
New York, New York 10047
(212) 488-4661

*North Carolina Department of Labor
4 West Edenton Street
Raleigh NC. 27601-1092
OSH info: 919 662- 4575
Consultative Services: 919 733-2360

Ohio Department of Industrial Relation
OSHA Survey Office
P.O. Box 12355
Columbus, Ohio
(405) 235-1447

Oklahoma Department Of Labor
Supplemental Data Division
315 North Broadway Place
Oklahoma City, Oklahoma 73105
(405) 235-1447

*Oregon Workers' Compensation Department
Research and Statistics Section
Labor and Industries Building
Salem, Oregon 97310
(503) 378-8254

Pennsylvania Department of Labor and Industry
7th and Forster Sts.
Labor and Industry Building
Harrisburg, Pennsylvania 17121
(717) 787-1918

*Puerto Rico Department of Labor and Human Resources
5050 Munoz Rivera Avenue
San Juan, Puerto Rico 00918
(809) 754-5339

Rhode Island Department of Labor
220 Elmwood Avenue
Providence, Rhode Island 02907
(401) 277-2731

*South Carolina Department of Labor
Division of Technical Support
P.O. Box 11329
Columbia, South Carolina 29211
(803) 734-9652

*Tennessee Department of Labor
501 Union Building, 2nd Floor
Nashville, Tennessee 37219
(615) 741-1748

Texas Department of Health
Division of Occupational Safety
1100 West 49th Street
Austin, Texas 78756
(512) 458-7287

*Utah Industrial Commission
OSH Statistical Section
160 East 300 South
Salt Lake City, Utah 84110-5800
(801) 530-6827

*Vermont Department of Labor and Industry
State Office Building
Montpelier, Vermont 05602
(802) 828- 2765

*Virgin Islands Department of Labor
P.O. Box 818
St. Thomas, Virgin Islands 00801
(809) 776-3700

*Virginia Department of Labor and Industry
205 North 4th Street
P.O. Box 12064
Richmond, Virginia 23241
(804) 786-2384

*State of Washington
Department of Labor and Industry
Division of Occupational Safety and Health
P.O. Box 2589
Olympia, Washington 98504
(206) 753-4013

West Virginia Department of Labor
OSH Project Director
Rm. 319, Bldg.3, Capitol Complex
1800 Washington Street East
Charleston, West Virginia 25305
(304) 348-7890

*Wyoming Department of Labor and Statistics
Herschler Building
Cheyenne, Wyoming 82002
(307) 777-6370

Accident Investigation

Associations and Organizations

FBI
SBI
U.S. Department of Labor OSHA
National Transportation Safety Board
National Fire Protection Association

Consultants

East Coast Accident Reconstruction & Investigations
323 W. Morgan Street
Raleigh, NC
(919) 821-0016

Corroon & Black of Raleigh, Inc.
100 Europa Drive
Raleigh, NC
(919) 832-2330

Electronic Resources

Internet info@webcrawler.comAccident Investigation

Manufacturers and Suppliers

J.J. Keller & Associates, Inc. (Compliance Audits)
1-800-327-6868

Publication Resources (Books, Government Publications, Journals, Periodicals, Newsletters, Reports)

National Underwriter (Periodical)
Occupational Hazards (Periodical)
Journal of Property Management (Journal)
Occupational Health & Safety (Periodical)
Business Insurance (Periodical)

Construction Industry Digest (Govt. Publication)
NFPA 921 Fire and Explosion Investigations (Publication)
Professional Safety

Electrical Safety

Associations and Organizations

National Fire Protection Association
U.S. Department of Labor OSHA
National Safety Council
American Society of Safety Engineers

Consultants

Ralph P. Cochrane, P.E.
P.O. Box 31246
Charlotte, NC 28231

Framatome Technologies
P.O. Box 10935
Lynchburg, VA 24506-0935
(804) 832-2484

Electronic Resources

National Fire Protection Association Electronic NEC
1 Batterymarch Park
P.O. Box 9146
Quincy MA 02269-9959
1-800-344-3555

Manufacturers and Suppliers

Bureau of Business Practice, Inc
24 Rope Ferry Road
Waterford, CT 06386
1-800-243-0876

National Fire Protection Association
1 Batterymarch Park
P.O. Box 9146
Quincy, MA 02269-9959
1-800-344-3555
NEC Code books

Publication Resources (Books, Government Publications, Journals, Periodicals, Newsletters, Reports)

Engineers Digest (Periodical)
Power Engineering (Periodical)
Electrical World (Periodical)

National Safety Council (Government Publication)
444 North Michigan Avenue
Chicago, Illinois 60611
1-800-621-7619, Ext. 6900

Emergency Response

Associations and Organizations

American National Red Cross
Federal Communication Commission (Emergency Broadcast System)
The Chemical Manufacturers Association (CMA)
Federal Emergency Management Agency

Electronic Resources

Internet info@Yahoo.comEmergency Response

Manufacturers and Suppliers

MSDS Sheets (All Manufacturers)
J.J. Keller & Associates, Inc. (Compliance Audits)
1-800-327-6868

Publication Resources (Books, Government Publications, Journals, Periodicals, Newsletters, Reports)

NFPA 101 Life Safety Code (Publication)
Seventeenth Edition of the Fire Protection Handbook (Publication)
Emergency Preparedness Manual (Publication APPA)
Emergency Response Guidebook (Published by U.S. DOT)

State and Federal Agencies

U.S. Department of Labor OSHA
U.S. Department of Transportation
The Environmental Protection Agency

Federal and State Regulatory Compliance

Consultants

Refer to the American Society of Safety Engineers' *National Directory of Safety Consultants*

Electronic Resources

Internet info@webcrawler.comRegulatory Compliance

Manufacturers and Suppliers

J.J. Keller & Associates, Inc. (Compliance Audits)
1-800-327-6868

Fire Prevention and Protection

Associations and Organizations

Industrial Risk Insurers
85 Woodland Street
Hartford, Connecticut 06102

National Fire Protection Association
1 Batterymarch Park
P.O. Box 9101
Quincy, MA 02269-9101
(617) 770-3000

Electronic Resources

Internet info@webcrawler.comFire Protection

Manufacturers and Suppliers

J.J. Keller & Associates, Inc. (Compliance Audits)
1-800-327-6868

ANSUL Incorporated
One Stanton Street
Marinette, WI 54143-2542
(715) 735-7411
Manufactures wheeled and hand-held portable fire extinguishers.

Justrite Manufacuring Co.
2454 Dempster Street
Des Plaines, IL 60016-5344
(708) 298-9250
Manufactures flammable-liquid storage cabinets, safety cans, and flammable waste disposal containers.

Bureau of Business Practice, Inc
24 Rope Ferry Road
Waterford, CT 06386
1-800-243-0876

Publication Resources (Books, Government Publications, Journals, Periodicals, Newsletters, Reports)

Bureau of Business Practice, Inc. (Books, Reports, and Forms)
24 Rope Ferry Road
Waterford, CT 06386
1-800-243-0876

National Safety Council (Government Publication)
444 North Michigan Avenue
Chicago, Illinois 60611
1-800-621-7619, Ext. 6900

Health and Safety Education and Training

Associations and Organizations

U.S. Department of Labor OSHA
Association of Physical Plant Administrators
National Safety Council
American Society of Safety Engineers
American Industrial Hygiene Association

Publication Resources (Books, Government Publications, Journals, Periodicals, Newsletters, Reports)

The Locomotive (Publication)
Keller's Official OSHA Safety Handbook (Publication)
Hazardous Chemicals "Right to Know Act" (Government Publication)
Hazardous Materials and Solid Waste Management (APPA Publication)
Regulatory Compliance for Facility Managers (APPA Publication)

Inspection and Audits

Associations and Organizations

U.S. Department of Labor OSHA
U.S. Environmental Protection Agency
Quality and Productivity Management Association

Consultants

Safety Management Services
10761 Oakhurst Drive
Rancho Cucamonga, CA 91730
(909) 989-8375

10761 Oakhurst Drive
Rancho Cucamonga, CA 91730
(909) 989-8375

Electronic Resources

Internet info@webcrawler.comSafety Inspections & Audits

Manufacturers and Suppliers

J.J. Keller & Associates, Inc. (Compliance Audits)
1-800-327-6868

Publication Resources (Books, Government Publications, Journals, Periodicals, Newsletters, Reports)

Quality Progress (Periodical)
Quality and Participation (Periodical)
Total Quality (Newsletter)

State and Federal Agencies

U.S. Department of Labor OSHA

References

Best's Safety Directory. 1995 ed., Vols. 1-2. Oldwick, NJ.: A. M. Best Company.

Fischer, C. A., and Schwartz, C. A. Eds. 1996. *Encyclopedia of associations: National organizations of the U. S.* 30th ed., Vol. 1. New York, NY.: Gale Research Inc.

Thurn, L. Ed. 1966. Encyclopedia of associations; International organizations, 30th ed. New York, NY.: Gale Research Inc.

Ulrich's international periodicals directory. 33rd ed., Vols. 2-3. 1994-1995. New Providence, NJ.: R. R. Bowker.

Appendix A

29 CFR 1910
OSHA General Industry Standards
Summary and Checklist

Table of Contents

Title 29 CFR 1910 OSHA General Industry Standards

Introduction

In 1970, the United States Government passed the Occupational Health and Safety Act PL-91593. This act was designed to ensure safe and healthy working conditions for men and women in the United States and any other areas falling into the jurisdiction of the United States. The Secretary of Labor is in charge of OSHA. OSHA enacts and enforces laws that pertain to safety in industry. OSHA also has the power to inspect places of business and to levy fines for violations of the codes or laws. The *Code of Federal Regulations* (CFR) is long and arduous and often confusing to the average person. Therefore, this is an overview of Title 29 CFR 1910, Occupational Safety and Health Standards for General Industry.

This document is not intended to be a replacement for 29 CFR 1910, but it can be used as a quick reference guide to 29 CFR 1910. Contractors or business persons look up a subpart in this appendix, get a brief overview of what it is about, and answer quick and easy questions. They can first go to this manual and decide which sections of the code to interpret. If they answer yes to a question in a subpart, then they know that they need to look at that subpart up in 29 CFR 1910 and make sure that their company complies with that standard.

Purpose

This document is to be used as an aid for compliance with federal safety and health regulations. The main purpose is to assist employers in determining the applicability of standards for specific industries. This is to be used only as a supplement to 29 CFR 1910 and cannot, in any way, provide specific information independently on compliance or regulatory law.

How to Use This Document

Part 1910 is broken down into subparts and each subpart into sections. Under each section there is a question or series of questions to determine applicability. If the answer is yes to any question under the section, then the section applies and the individual needs to be referred to the appropriate subpart and section for clarification in 29 CFR 1910. If there is doubt about whether a company is in compliance on any portion of 29 CFR 1910, it is important to confirm that fact by referring to the standard.

Subpart A - General

This subpart explains the purpose and scope, definitions, petitions, amendments, and applicability of all the standards in 29 CFR 1910. Standards will be adopted as necessary by the Secretary to make a healthier and safer workplace. It explains each individual's role from the Secretary of Labor to that of the employee, as well as other terms used throughout 29 CFR 1910. Any interested person may petition in writing to the Assistant Secretary of Labor for the issuance, amendment, or repeal of a standard. It places the Assistant of the Secretary of Labor at the same level of authority as the Secretary of Labor. It lists all the territories that the standards apply to but excludes governmental agencies.

Subpart B - Adoption and Extension of Established Federal Standards

This subpart adopts and extends the applicability of established Federal standards to every employer, employee, and place of employment covered in the Act. Only standards relating to safety or health are adopted into this Act. This also pertains to any facility engaging in construction, alterations, or repair, including painting and decorating. Additionally, it pertains to facilities with employees engaged in ship repair, ship breaking, or that use craft as a means of transportation on water or a related employment. Companies employing people aboard a vessel on navigable waters of the United States and employees engaged in longshoring operations, i.e. the moving or handling of ship's cargo, stores into, on, in, or out of any vessel are also covered by the Act. Facilities which expose any employee to asbestos, tremolite, anthophyllite actinolite dust, vinyl chloride, inorganic arsenic, lead, ethylene oxide, acrylonitrile, formaldehyde, and cadmium are subject to the Act, as well.

Subpart C - General Safety and Health Provisions

The purpose of this subpart is to provide employees and their designated representatives a right of access to relevant exposure and medical records. It also provides the representatives of the Assistant Secretary a right of access to these records in order to fulfill responsibilities under the Occupational Safety and Health Act.

Subpart D - Walking - Working Surfaces

This subpart addresses the requirements for maintaining walking and working surfaces. Subpart D applies to all permanent places of employment, except where domestic, mining, or agricultural work only is performed. It contains regulations pertaining to housekeeping, aisles and passageways, ladders, scaffolding, railing, and working surfaces, including mobile surfaces. The section also contains information on guarding floor and wall openings and holes, fixed industrial stairs, portable wood ladders, portable metal ladders, and fixed ladders. Safety requirements for scaffolding, manually propelled mobile ladder stands and scaffolds (towers), and other working surfaces such as dockboards (bridge plates), forging machine areas, wooden platforms, veneer machinery space requirements, and walkways adjacent to large steam vats are presented in this subpart.

Subpart E - Means of Egress

This subpart deals specifically with Means of Egress, defined by the standard as "... a continuous and unobstructed way of exit travel from any point in a building or structure to a public way and consists of three separate and distinct parts: the way of exit access, the exit, and the way of exit discharge." The standard addresses each of these items to include general requirements, the various items that make up an exit, specific physical requirements for an exit, and the number of exits required. This subpart also contains the general fundamental requirements essential to providing a safe means of egress from fire and like emergencies. The section has information on definitions, general requirements, means of egress, and employee emergency plans and fire prevention plans. It discusses the minimum requirements for the plans, alarm systems to be used, training requirements, and maintenance of equipment used under the plans.

Subpart F - Powered Platforms, Manlifts, and Vehicle - Mounted Work Platforms

This section covers powered platform installations permanently dedicated to interior or exterior building maintenance of a specific structure or group of structures. It does not apply to suspended scaffolds (swinging scaffolds) used to service buildings on a temporary basis and covered under Subpart D of this part, nor suspended scaffolds used for construction work and covered under Subpart L of 29 CFR Part 1926. This section applies to all permanent installations completed after July 23, 1990, and contains information on powered platforms for building maintenance. Building maintenance includes, but is not limited to, such tasks as window cleaning, caulking, metal polishing, and reglazing. This section goes on to address such items as building owner and employer responsibilities associated with this equipment; specific make-up of the equipment, engineering, and the design of the equipment; and its inspection and maintenance requirements. It also addresses employee training, personal protective equipment requirements, and general housekeeping items.

In addition, this section specifically addresses the requirements for vehicle-mounted elevating and rotating work platforms. It specifically states: "Unless otherwise provided in this section, aerial devices (aerial lifts) acquired on or after July 1, 1975, shall be designed and constructed in conformance with the applicable requirements of the American National Standard for Vehicle-mounted Elevating and Rotating Work Platforms, ANSI A92.2-1969, including appendix." The employee training, personal protective equipment required, boom operation around electrical hazards, and inspection and maintenance of equipment are also discussed.

Subpart G - Occupational Health and Environmental Control

The standards in subpart G deal with air quality, noise exposure exceeding eighty- five decibels, and radiation exposure in the workplace. This includes facilities that use abrasive blasting; facilities that have spray booths, spray rooms, or open surface tanks used for cleaning; facilities with grinding, polishing, and buffing operations; and facilities that have ionizing radiation such as alpha rays, beta rays, gamma rays, x-rays, and other atomic particles, or radioactive materials such as nonionizing radiation that may be present in the facility.

Subpart H - Hazardous Materials

Subpart H contains information on compressed gasses, acetylene, hydrogen, oxygen, nitrous oxide, flammable and combustible liquids, spray finishing using flammable and combustible materials, dip tanks using flammable or combustible liquids, explosive and blasting agents, storage and handling of liquid petroleum gasses, and storage and handling of anhydrous ammonia. This section also covers process-safety management requirements of highly hazardous chemicals and hazardous-waste operations and emergency response.

Subpart I - Personal Protective Equipment

This subpart provides the general requirements for the minimum personal protective equipment to be used by the employees in the performance of their tasks for protection against recognized workplace hazards. The section specifically states: "Protective equipment including personal protective equipment for eyes, face, head, and extremities; protective clothing; respiratory devices; and protective shields and barriers; shall be provided, used, and maintained in a sanitary and reliable condition wherever it is necessary by reason of hazards of processes or environment, chemical hazards, radiological hazards, or mechanical irritants encountered in a manner capable of causing injury or impairment in the function of any part of the body through absorption, inhalation, or physical contact" (Personal Protective Equipment, 1993, p. 205). It goes on to address employee-owned equipment; design, hazard assessment, and equipment selection; employee training; and inspection and maintenance of personal protection, foot protection, and hand protection.

Subpart J - General Environmental Control

This section specifically applies to permanent places of employment. It addresses such items as toilet facilities, including numbers and type; washing facilities; sanitary food storage; and food handling. It addresses temporary labor camps, safety color- coding for marking physical hazards, and specifications for accident prevention signs and tags. Two additional items specifically addressed by this section and of considerable importance are permit-required confined spaces and the control of hazardous energy (lockout/tagout).

The permit-required confined spaces standard defines a "confined space;" lists specific employee training requirements, equipment, and personal protective equipment required prior to entry; and communications and emergency procedures. Appendix A to the 1910 standard also includes an informative Permit-Required Confined Space Decision Flow Chart, as well as an example of a Confined Space Entry Permit.

The standard for the control of hazardous energy (lockout/tagout) addresses "the servicing and maintenance of machines and equipment in which the unexpected energization or start-up of the machines or equipment or release of stored energy could cause injury to employees. This standard establishes minimum performance requirements for the control of such hazardous energy" (Control of Hazardous Energy, 1993, p. 225). It further spells out the requirements of the standard including all aspects of training and personal protective equipment, as well as methods to meet the standards.

Subpart K - Medical and First Aid

The purpose of Subpart K - Medical and First Aid is to provide the employee with readily available medical consultation on matters related to on-plant health. This service may be provided by on- or off-site personnel. First aid supplies shall also be readily available.

Subpart L - Fire Protection

This subpart contains requirements for fire brigades, all portable- and fixed-fire suppression equipment, fire detection systems, and fire or employee alarm systems installed to meet the fire protection requirements of 29 CFR 1910. It contains requirements for the organization, training, and personal protective equipment of fire brigades whenever they are established by an employer. These requirements apply to fire brigades, industrial fire departments, and private or contractual fire departments. Personal protective equipment requirements apply only to members of fire brigades, performing interior structural fire fighting. The requirements of this section do not apply to airport crash rescue or forest fire fighting operations. In addition, this subpart establishes the requirements for the placement, use, maintenance, and testing of portable fire extinguishers provided for use by employees, as well as the requirements for all automatic sprinkler systems installed to meet a particular OSHA standard.

Subpart M - Compressed-Gas and Compressed-Air Equipment

This section applies to compressed-air receivers and other equipment used in providing and utilizing compressed air for performing operations such as cleaning, drilling, hoisting, and chipping. However, this section does not deal with the special problems created by using compressed air to convey materials, nor the problems created when work is performed in compressed-air environments such as in tunnels and caissons. This section is not intended to apply to compressed-air machinery and equipment used or transportation vehicles such as steam railroad cars, electric railway cars, and automotive equipment.

Subpart N - Materials Handling and Storage

This section is a general overview that covers the handling and storage of materials. The main concern of this subpart is the movement of materials on-site. Its four sections are devoted to material handling devices. These devices consist of industrial trucks, overhead and gantry cranes, crawler and truck cranes, derricks, and slings used to connect the loads. The use of helicopters to lift materials is covered, but more stringent FAA regulations exist that cover this in detail. The servicing of single and multi-piece rims are also covered in this subpart.

The industrial trucks section covers the classifications of trucks and designated areas where a truck can be used. It also describes the required inspections and maintenance actions for those vehicles. Safe operation procedures are also covered in this section.

The same material is covered for cranes and derricks. Loads and proper lifting procedures are also covered. Required inspections are outlined and procedures to ensure that a non-functioning machine is marked and safe.

Procedures for keeping and using slings is also covered in this section. It describes the proper sizes for loads as well as safe hook-up procedures. Inspection requirements are stated and required markings are discussed.

The servicing of single-piece and multi-piece rim wheels is covered for those pieces of equipment that are not covered by the Construction Safety Standards, the Agriculture Standards, the Shipyard Standards, or the Longshoring Standards.

Subpart O - Machinery and Machine Guarding

Subpart O covers the machine guarding for any equipment that exposes the employee to a hazard during use due to exposed moving or rotating parts, generally speaking, this covers any device that has an exposed point of operation. This subpart covers guards for woodworking machinery, abrasive-wheel machinery, mills and calendars in the plastics industries, mechanical presses, forging machines, and mechanical power-transmission apparatus.

The woodworking section covers the parts that must be guarded and the type of guards that must be used, while the abrasive-wheel section describes the amount of wheel which can be exposed for the various types of abrasive-grinding equipment. Proper mounting is described for the various wheels. Charts of the minimum dimensions for abrasive wheels and their safety guards are included.

Mechanical-power presses are required to have switches and brakes that are operator proof. The size of some presses makes it imperative that they are switch- guarded as well as mechanically guarded. This section describes what actions must be taken to produce the most failsafe device possible. Minimum braking action and the proper brakes that must be installed on presses of differing specifications are outlined. Criteria for the use of presence-sensing devices and their proper utilization is described.

The safety criteria for a forging operation is the next section. Operation, inspection, and maintenance requirements are covered. The requirements for venting special processes and equipment that needs to be guarded are included.

The mechanical-power-transmission apparatus section covers all belts, pulleys, and conveyors that are used in industry. It describes which ones need to be guarded. For specific applications, guidance is given for preferred actions.

Subpart P - Hand and Portable Powered Tools and Other Hand-held Equipment

This section describes the required guarding, inspections, and maintenance requirements for hand and portable tools. It includes tools employees may own and use on the job. Lawn mowers and other internal-combustion-engine-powered machines are included in this section. The proper procedures for using portable jacks are also covered in this section.

Subpart Q - Welding, Cutting and Brazing

Subpart Q covers the use and installation of electric or gas welding, cutting, and brazing equipment. It covers the different types of welding and the specific safety needs of each. This subpart also regulates the use of oxygen-fuel gas welding and cutting, arc welding and cutting, and resistance welding. Basic fire precautions are included in this subpart, not detailed regulations, which are covered in NFPA standard 51B, 1962. Maintenance and safety requirements are included in this subpart, including welding in confined spaces and proper ventilation.

Subpart R - Special Industries

This section applies to special industries as defined by OSHA. These special industries include paper, pulp and paperboard mills, textile mills, bakeries, laundries, and sawmills. Other industries covered are pulpwood logging; agricultural operations; telecommunications; electric-power generation, transmission, and distribution; and finally, grain-handling facilities. These special industries are trades that perform specific tasks that are very distinct and, in turn, set standards for their own industries.

Subpart S - Electrical

This subpart addresses the electrical safety requirements for the safeguarding of employees in their workplace. It is divided into sections dealing with the general requirements of wiring design, protection of specific-purpose equipment and installations. In addition, this subpart covers locations, wiring methods, wiring components, and training and equipment used.

Subpart T - Commercial Diving Operations

This subpart applies to every place of employment within territory the of the United States, or within any state, the District of Columbia, the Commonwealth of Puerto Rico, the Virgin Islands, American Samoa, Guam, the trust Territory of the Pacific Islands, Wake Island, Johnston Island, the Canal Zone, or within the Outer Continental Shelf lands as defined in the Outer Continental Shelf Lands Act, where diving and related support operations are performed.

This standard applies to diving and related support operations conducted in connection with all types of work and employments including general industry, construction, ship repairing, shipbuilding, shipbreaking, and longshoring. It covers the personnel requirements, general operations procedures, specific operations procedures, equipment procedures and requirements, and recordkeeping for diving operations.

Subparts U-Y (Reserved)

Subpart Z - Toxic and Hazardous Substances

The purpose of this subpart is to protect employees from exposure to toxic and hazardous substances in the workplace. It covers the permissible Exposure Limits (PEL) for all air contaminants including all gases, vapors, and dusts. Some of the contaminants covered under this subpart include asbestos, coal tar pitch volatiles, vinyl chloride, inorganic arsenic, lead, cadmium, benzene, coke-oven emissions, blood-borne pathogens, cotton dust, ethylene oxide, and formaldehyde.

In addition, this subpart is Hazard Communication and Occupational Exposure to Hazardous Chemicals in the Laboratory. These regulations attempt to ensure worker awareness and knowledge of hazardous materials. Material Safety Data Sheets, container labeling, and employee-training requirements are specified in this subpart of the standards.

Subpart Z tables for the Permissible Exposure Limits of all air contaminants are published in this portion of the standards. Compliance requirements for hazardous materials found in these tables are presented in terms of eight-hour time weighted averages (TWA), maximum short-term exposure limits (STELs) and ceiling limits (C).

CHECKLIST FOR SUBPARTS A THROUGH Z

Subpart A - General
Purpose, scope, and definitions only in this subpart.

Subpart B - Adoption and Extension of Established Federal Standards
- Does this facility engage in construction, alterations, and/or repair, including painting and decorating?
- Does this facility have employees engaged in ship repair, ship breaking, shipbuilding, or use craft as a means of transportation on water or a related employment?
- Does the company employ people aboard a vessel on the navigable waters of the United States?
- Does the company have employees engaged in longshoring operations i.e., the moving or handling of ship's cargo, stores or gear into, on, in, or out of any vessel?
- Is the facility a marine terminal such as a pier, dock, or wharf?
- Does the facility expose any employee to asbestos, vinyl chloride, inorganic arsenic, lead, ethylene oxide, acrylonitrile, formaldehyde, or cadmium?

Subpart C - General Safety and Health Provisions
- Does the facility compile and retain employee medical records?

Subpart D - Walking - Working Surfaces
- Does the facility have passageways, storerooms, or service rooms?
- Does the facility have mechanical handling equipment which must be operated along aisles and passageways?
- Does the facility have open tanks, pits, vats, ditches, etc., which need covers and guardrails to protect employees?
- Is this facility a permanent place of employment?
- Does the facility have stairway, ladderway, hatchway, and chute floor openings?
- Does the facility have any wall openings, manhole floor openings, or temporary floor openings?
- Does the facility have fixed stairways inside or out; around machinery, tanks, and other equipment; and stairs leading to or from floors, platforms, or pits?
- Does the facility have ladders?

- Is fixed or portable scaffolding used at the facility?
- Are dockboards or bridge plates used at the facility?
- Does the facility have forging machines?
- Are steam vats used at the place of employment?

Subpart E - Means of Egress

- Does the facility have emergency exits?
- Does the facility have the required written emergency action plan and fire-prevention plan?
- Has the facility properly trained a sufficient number of persons to assist in an orderly emergency evacuation?

Subpart F - Powered Platforms, Manlifts, and Vehicle - Mounted Work Platforms

- Are powered platforms of any type in use at the facility?
- Is training of employees in the operation and inspection of working platforms performed by a competent person?

Subpart G - Occupational Health and Environmental Control

- Does the facility use abrasive blasting--a solid substance used in blasting operations?
- Does the facility have spray booths, spray rooms, or open surface tanks used for cleaning?
- Does the facility have grinding, polishing, and buffing operations?
- Does the facility have noise levels exceeding eighty-five decibels?
- Is ionizing radiation such as alpha rays, beta rays, gamma rays, x-rays, and other atomic particles present in the facility?
- Are radioactive materials present in the facility?
- Is nonionizing radiation such as electromagnetic radiation which includes both the radio and microwave frequency regions present in the facility?

Subpart H- Hazardous Materials

- Does the facility or work area have compressed-gas cylinders at the location?
- Is acetylene transferred through a piped system?
- Does the facility generate or fill cylinders with acetylene?
- Is hydrogen delivered, stored, or discharged at the facility?
- Does the facility have a gaseous hydrogen system in place?
- Does the facility have a liquefied hydrogen system in place?
- Does the facility have hydrogen containers?
- Does the facility have a bulk oxygen system?

- Is nitrous oxide used in the facility?
- Does the facility use flammable and combustible liquids in forms such as aerosol cans and atmospheric tanks?
- Does the facility have underground storage tanks for dispersement from fixed equipment into the fuel tanks of motor vehicles such as service stations?
- Are aerated solid powders used as a coating material used at the facility?
- Are spray booths spray areas, waterwash spray booth, or dry spray booth used at the facility?
- Does the facility have a tank, vat, or container of flammable or combustible liquid in which articles or materials are immersed for the purpose of coating, finishing, treating, or similar processes?
- Are any of the liquids in the dip tank heated?
- Are rags used or waste generated in connection with dipping operations?
- Does the facility have any vapor areas where dangerous quantities of flammable vapors can build up during operation or shutdown periods?
- Does the facility have or use class A, B, or C explosives and blasting agents?
- Are pyrotechnics, which includes fireworks, manufactured or stored at the facility?
- Does the company store, transport, or handle liquefied petroleum gasses in containers such as tanks, cylinders, or drums?
- Does the company store, transport, or handle anhydrous ammonia (excluding manufacturing and plants where ammonia is used solely as a refrigerant), in containers such as tanks, cylinders, or drums?
- Does the operation have the potential for producing a catastrophic release of toxic, reactive, flammable, or explosive chemicals?
- Does the process involve a flammable liquid or gas on-site in one location in a quantity of 10,000 lbs. or more?
- Does the company have employees who handle or may come in contact with hazardous waste or substances and chemicals?
- Do employees such as fire fighters or mutual aid groups respond to emergencies which are likely to result in an uncontrolled release of a hazardous substance?
- Are any employees designated by the employer as a Hazardous Materials Response Team (HAZMAT)?

Subpart I - Personal Protective Equipment
- Does the facility have any hazardous process, environmental hazards, chemical hazards, radiological hazards, or mechanical irritants which

could cause injury to any part of the body by physical contact, breathing, or absorption?

- Does the facility require the employees to provide their own protective equipment?
- Are any employees exposed to eye or face hazards from flying particles, molten metal, liquid chemicals, acids, caustic liquids, chemical gases or vapors, or potentially injurious light radiation?
- Are employees exposed to hazards from flying objects?
- Do any employees wear prescription lenses while engaged in operations that involve eye hazards?
- Are employees exposed to any welding activities?
- Is there any exposure to occupational diseases caused by breathing contaminated air, whether in the form of dusts, fogs, fumes, mists, gases, smokes, sprays, or vapors?
- Are respirators provided when the equipment is necessary to protect the health of the employee?
- Do employees receive training in the use of respiratory protection?
- Are any employees working in areas where there is a potential for injury to the head from falling objects?
- Is there any possible exposure to electrical hazards?
- Are employees working in areas where there is a danger of foot injuries due to falling or rolling objects, objects piercing the sole, or where employee's feet are exposed to electrical hazards?
- Are employees exposed to electrical hazards?
- Are employee's hands exposed to hazards such as skin absorption, heat changes, or burns?

Subpart J - General Environmental Control

- Are there any physical hazards that will need to be marked with appropriate colors?
- Are there signs and symbols needed to indicate and define specific hazards?
- Are employees exposed to the hazards of entry into permit-required confined spaces?
- Does the facility have any machinery that could *unexpectedly* energize, start-up, or release stored energy and cause injury to employees?
- Are employees required to remove or bypass a guard or other safety device for any reason?
- Are employees required to place any part of the body into an area on a machine where work is actually performed, or where an associated danger zone exists during machine operation?

Subpart K - Medical and First Aid
- Does the facility provide medical equipment and personnel for advice and consultation on medical matters?

Subpart L - Fire Protection
- Are fire protection methods in place at the facility?
- Does the facility have a fire brigade?
- Does the facility use industrial fire departments or private or contractual fire departments?
- Are fire extinguishing systems present at the facility?
- Does the facility have any automatic fire detection systems?
- Does the facility have any emergency employee alarms?

Subpart M - Compressed-Gas and Compressed-Air Equipment
- Does the facility have a compressed-air receiver?
- Does the facility have equipment that provides compressed air?

Subpart N - Materials Handling and Storage
- Does the facility store materials or have storage areas?
- Does the facility receive materials by rail or have a railway.
- Does the facility service single- or multi-piece rims for trucks, tractors, trailers, buses, and off-road machines? This does not apply to construction, agriculture, or longshoring.
- Does the facility utilize fork trucks, motorized hand trucks, or other specialized industrial trucks that are powered by electric motor or internal-combustion engines?
- Does the facility operate cranes of any type?
- Does the facility utilize derricks?
- Does the facility operate a helicopter to lift, move, or otherwise handle materials?
- Does the facility use any rope, chain, cable, or strap for slings that connect a load to a material-handling device for the purpose of hoisting or pulling?

Subpart O - Machinery and Machine Guarding
- Does the facility operate machinery that has any of the following characteristics including but not limited to point of operation area, ingoing nip points, rotating parts, exposed, or flying chips and sparks that create hazards to employees?
- Does the facility operate woodworking machinery?

- Does the facility operate any abrasive-wheel machinery?
- Is the facility in the rubber or plastics industries?
- If the above is true, does the facility operate mills and calendars?
- Does the facility operate mechanical-power presses, excluding the following machines: press brakes, hydraulic and pneumatic-power presses, forging presses and hammers, bulldozers, hot-bedding and hot-metal presses, riveting machines and similar types of fastener applicators?
- Does the facility have a forging or die shop that uses lead casts or other uses of lead?
- Does the facility have power-transmission belts?

Subpart P - Hand- and Portable-Power Tools and Other Hand-held Equipment

- Does the facility or its employees use hand tools that may pose a hazard to the employee if not maintained properly?
- Does the facility use compressed air for cleaning purposes?
- Does the facility use portable power tools that have an exposed point of operation that creates a hazard to the employee?
- Does the facility have any internal-combustion- engine-powered portable tools?
- Does the facility use portable jacks?
- Does the facility use portable blast cleaners?

Subpart Q - Welding, Cutting and Brazing

- Does the work require the use of electric or gas welding and cutting equipment?
- Is the work to be done near a fire hazard?
- Is a mixture of fuel gas and air or oxygen to be used for welding or cutting?
- Are compressed-air or oxygen cylinders stored or transported?
- Are pressure generators utilized?
- Is arc welding and cutting equipment used?
- Is resistance welding utilized?

Subpart R - Special Industries

- Does the work involve the manufacturing and converting of pulp, paper, and paperboard?
- Are forklift trucks, hand/power trucks, cranes, ships, trucks, or railroad cars used in the moving of materials for manufacturing purposes?
- Does the work involve the design, installation, processes, operation, and maintenance of textile machinery (except synthetic fiber)?

- Is steam used in the processes?
- Does the work involve the design, installation, operation, and maintenance of machinery and equipment used within a bakery?
- Is flour-handling equipment used?
- Are steam pipes used?
- Does the laundry equipment utilize moving parts at the point of operation?
- Are chemicals or flammable liquids used in the processes throughout the facility?
- Are vehicles used to transport the material within the area?
- Are kilns utilized on the facility? Plywood, cooperage, and veneer manufacture are excluded from this standard.
- Does the work involve the normal operations included in the logging of pulpwood (excludes sawlogs)?
- Does the work involve the installation and processes performed at telecommunications centers and at telecommunications field installations?
- Is the work performed around energized power lines, batteries, or hazardous materials?
- Does the task require working in confined space or manhole worksites?
- Does the enterprise involve working with the operation and maintenance of electric-power generation, control transformation, transmission, and distribution lines and equipment?
- Does the work require the inspection of wood poles?
- Does the work require the trimming of trees around power lines?
- Does the enterprise require the operation of grain elevators, feed mills, flour mills, rice mills, dust-pelletizing plants, dry corn mills, soybean flaking operations, and dry grinding operations of soy cake?
- Is welding, cutting, or brazing to be performed?

Subpart S - Electrical
- Does the work include the installation and examination of electrical equipment used to provide electrical power and light for employee workplaces?
- Are all conductors and equipment approved?
- Are hoists or splices used during installation?
- Is there flammable or combustible hazards nearby?
- Is equipment used that requires markings?
- Are conductors utilized above ground that are near walkways, working areas, crossover roads or railways, near or over the tops of buildings, or near moving mechanisms such as cranes or booms?
- Are metal raceways, cable armor, and other metal enclosures used?
- Is temporary wiring done?

- Does wiring run through walls, floors, wood cross members, etc.?
- Does the work involve the installation or maintenance of electrical signs or directional lighting?
- Are cranes or hoists used to conduct the work?
- Are elevators, dumbwaiters, escalators, and moving parts used?
- Is the electrical work to be performed in an area that contains flammable vapors, liquids, or gases, or combustible dusts or fibers?
- Are there electrical systems over 600 volts?
- Is the electrical system for fixed installations or mobile equipment?
- Is the electrical system for emergency power supply?
- Is the employee to perform electrical work qualified or unqualified? The employee's risk of shock is not reduced by the electrical installation standards provided by the previous sections.
- Do the employees receive regular documented training?
- Is the work to be performed near energized or de-energized electrical equipment?
- Is corded portable electrical equipment used?
- Are employees working near areas where there is the potential for electrical hazard?

Subpart T - Commercial Diving Operations

- Is the diving work related to general industry, construction, shipbuilding, ship repairing, ship breaking, or longshoring?
- Are dive teams experienced and properly trained?
- Has the employer developed and maintained a safe-practices manual and made this manual available at the dive locations for all dive team members?
- Does the manual contain emergency and safety procedures, equipment procedures, etc.?
- Does the employer comply with the necessary safety and emergency requirements prior to each diving operation?
- Does the employer follow requirements that are applicable during all dives?
- Does the employer observe the required post-dive procedures?
- Are employees engaged in SCUBA diving?
- Does the employer engage surface-supplied air diving?
- Are employees engaged in mixed-gas diving?
- Are employers engaged in diving operations lifeboating?

Subparts U-Y (Reserved)

Subpart Z - Toxic and Hazardous Substances

* Does the facility have any air contaminates listed in tables Z-1, Z-2, or Z-3?
* Does the facility contain any asbestos in any form?
* Does the facility have any of the following: 4-nitrobiphenyl, alpa-Napthylamine, Methyl Chloromethyl Ether, 3,3'-Dichlorobenzidine, bis-Chlormethyl Ether, beta-Naphthylamine, Benzidine, 4-Aminodiphenyl, Ethyleneimine, beta-Propiolactone, 2-Acetylaminofluorene, 4-Dimethylaminoazobenzene, N-Nitrosodimethylamine, and Vinyl Chloride?
* Are employees at risk of exposure to inorganic arsenic, lead, cadmium or cadmium compounds, benzene, coke-oven emissions, occupational exposure to blood or other potentially infectious materials, or cotton dust?
* Does the facility manufacture yarn, perform slashing or weaving, or handle waste for textile operations?
* Are employees potentially exposed to 1,2 dibromo-3-chloropropane, Acrylonitrile, Ethylene Oxide, Formaldehyde, or Methylenedianiline?
* Does the facility manufacture or import hazardous chemicals?
* Does this facility use hazardous chemicals in a laboratory?

Reference

United States Department of Labor. 1993. *Occupational safety and health standards for general industry.* Washington, DC: U.S. Government Printing Office.

Appendix B

29 CFR 1926
OSHA Construction Standards
Summary and Checklist

Table of Contents

29 CFR 1926 OSHA CONSTRUCTION STANDARDS

Introduction

The Code of Federal Regulations Title 29 part 1926, covers compliance safety issues for construction companies. This document is a consolidation of OSHA standards which are applicable to the construction industry. In publishing this standard, every effort was made to identify all the applicable General Industry standards from 29 CFR 1910. It should be noted that, under certain circumstances, other parts of 29 CFR 1910 may also be relevant. This is a short summary of the requirements needed to establish a safe work environment and should be used as a guide for compliance for companies, facilities, and other businesses.

The appendix is arranged so that 29 CFR 1926 standards, subparts A through X appear first. This is followed by General Industry standards 29 CFR 1910 pertaining to construction, a subject index, and a listing of material approved for incorporation by reference. *Note:* Incorporation by reference is an established procedure whereby federal agencies meet publishing requirements by referring materials published elsewhere in the *Federal Register*

Subpart A - General

The first section of Subpart A is a general-purpose statement and an explanation of related policies. Subpart A is a general overview of the purpose, variances, inspections, and rules of practice used throughout the following subparts. In addition, the right of entry is addressed. This right of entry states that any contract subject to section 107, must allow the Secretary of Labor or a representative the right of entry. It is the right of the Secretary of Labor to conduct inspections or investigations to ensure compliance. The rules of practice for administrative adjudications for the enforcement of safety and health are also mentioned.

Subpart B - General Interpretations

Subpart B contains an interpretation and the applicability of construction health and safety standards to work performed under contract to government agencies. This subpart contains the general rules of the Secretary of Labor which state that no contractor shall require any laborer to work in conditions that are unsanitary, hazardous, or dangerous to a person's safety and health. Section 107 of the Contract Work Hours and Safety Standards Act is addressed. As stated in section 107, a contract must be one which is entered into under statute that is subject to Reorganization Plan Number 14 and is for construction.

Further details on the Reorganization Plan Number 14 illustrates that appropriate regulations with respect to labor standards are met. There are 58 statutes listed in the Reorganization Plan Number 14. Rental Housing, Federal-Aided Highways, and Defense Housing are some statutes listed in the Plan. The intent of the subpart is to ensure a safe working environment at any federally assisted project. In addition, there is a discussion of the definition of contractor and subcontractor and their respective responsibilities. Other issues mentioned in this subpart were basic rules followed by contractors and how to use these in relation to construction. Issues like toilets or first aid are the responsibility of the prime contractor but can be delegated to other subcontractors. However, the prime contractor is never relieved of the responsibility. In either situation, all regulations must be followed under the contract.

Subpart C - General Safety and Health

Subpart C covers the contractor requirements for general safety and health. It states that no contractor or subcontractor may require a worker to perform hazardous work or perform in adverse conditions. A detailed discussion of contractor/employer responsibilities is broken into eleven broad areas that includes:

- General Safety and Health Provisions
- Safety Training and Education
- Recording and Reporting of Injuries
- First Aid and Medical Attention
- Fire Protection and Prevention
- Illumination
- Sanitation
- Personal Protective Equipment (PPE)
- Acceptable Certification
- Shipbuilding and Ship Repair

In addition to the above referenced categories, there are other issues included in this subpart. For example, the contractor must also start and maintain accident prevention programs. Access to employee exposure and medical records is a key issue in this subpart.

Trade secrets are a big concern among employers and OSHA has allowed for them. Employers do not have to disclose secret formulas of dosages to inspectors or doctors in emergency situations if alternative information can be provided. However, under very special circumstances doctors may be authorized, following the special restraints of OSHA, to use the trade secrets in assisting them in their medical diagnoses. This procedure is governed by a confidentiality agreement of OSHA.

Further issues mentioned are means of exit. In general, all exits must be kept clear of any obstructions or impediments. Emergency escape procedures are also mandated by OSHA to insure an effective and safe evacuation route. In light of emergency actions, alarm systems and detail planning routes should be utilized by all employers.

Subpart D - Occupational Health and Environmental Controls

In Subpart D, a diverse group of issues are discussed. The following is a list of some of the highlights mentioned in Subpart B.

- Medical services is the first mentioned. It states the employer shall ensure the availability of medical personnel for advice or consultation on the issues surrounding occupational health. For example, first-aid supplies shall be made available to employees.
- The second issue discussed is the number of toilet seats in a facility. This is designated by the number of employees at the site. Twenty or fewer people require at least one seat, twenty or more people have one seat and one urinal per forty employees, and two-hundred or more people have one seat and one urinal for every fifty employees. Proper washing facilities must also be present for those employees working with harmful substances.
- In the area of noise exposure, a chart lists the many combinations of time and noise levels. Ionizing and non-ionizing radiation requirements are stated, as well as all the different precautions needed.
- Approximately 500 different substances are listed in a Threshold Limit Value table showing the acceptable limits of airborne particles. There are many examples on the proper installation of exhaust fans on grinders and belt sanders. These exhaust fans help to disperse and filter the particles for compliance.
- Asbestos and its removal is another topic discussed in Subpart D. This section applies to nearly any work with or in close proximity to asbestos. An extensive program must be maintained while working with asbestos. Airborne particles have to be stringently monitored. Even the laundering of clothes used while working with asbestos must be controlled.
- Another main point illustrated in this subpart is Process Safety Management. This is a program used by those who frequently utilize very dangerous chemicals and other harmful substances. Process Safety Management is a system that helps to ensure a safe working environment for

employees by using proper procedure and handling techniques with these materials.

• The final portion of this subpart deals with hazard communications. "This occupational safety and health standard is intended to address comprehensively the issue of evaluating the potential hazards of chemicals and communicating information concerning hazards and appropriate protective measures to employees..." (1926.59 (a) (2)).

Subpart E - Personal Protective and Lifesaving Equipment

This subpart deals with all the protective equipment that employers must supply to their employees for protection in the workplace. This includes personal protective equipment for the eyes, face, head, and extremities. In addition, protective clothing, respiratory devices, protective shields, and barriers shall be provided, used, and maintained in a sanitary and reliable condition wherever it is necessary because of hazards.

The employer is responsible for providing the employees with all the protective equipment that is needed for a particular job. They must also train the employees on how to properly use the equipment. Subpart E tells the employer what equipment is to be used for specific emergencies and where the equipment should be placed to ensure that all employees have knowledge of its location.

Hard hats are to be worn in areas where there is any possible danger of head injury from an impact of falling objects. Eye and face protection is to be used when machines or operations have the possibility of causing eye or face injury to the employee. This part also deals with the proper breathing apparatus needed in areas that are hazardous to the employees' health.

Subpart F - Fire Protection

Subpart F is concerned with fire protection, fire prevention, flammable and combustible liquids, liquefied petroleum, and temporary heating devices. Employers must have fire extinguishers in areas where fire is likely to occur. They must have the proper fire extinguisher for the hazard and the extinguishers must be approved by a nationally recognized testing laboratory. The facility must have either a temporary or permanent water supply that has enough pressure to put out a fire. A training course in proper fire extinguisher operation should be given to the employees to show how to operate fire extinguishers properly.

Flammable and combustible liquids should be stored in a container away from any activity which might cause an explosion. Most liquids must be stored in an approved room, and if they are stored outside of that room, it should be in quantities less than 25 gallons. In addition, this subpart informs the employer how close the employee must have the container for refilling operations and on moving containers from one area to another.

Tanks storage is broken down into different areas such as welding tank storage and the different types of tanks. Specifications are also given in this subpart on how the tanks should be built. Each tank is required to have certain values that regulate the pressure in the tanks. Tanks are to be stamped or marked with the material the tank is holding. Finally, inspection of these tanks is required in order to prevent an accident from happening.

Fire detection systems are another way of preventing a fire from happening. The employer must inspect this equipment on a timely basis to make sure that it is working properly and is in compliance with the OSHA standard. The detection system must be located in the specified area and should be labeled to note its location.

Subpart G - Signs, Signals, and Barricades

This subpart details the signs required when hazards are present and the notifications required when an employee is near that hazard. There are six main signs, danger, caution, exit, safety instruction, directional, and traffic signs, that are used on a construction site. The subpart also states what color each sign must be and the size requirements for every sign.

Accident prevention tags are used to warn the employee of an existing hazard such as defective tools or equipment. Flagmen should be used in areas such as streets and highways to provide direction for vehicles to travel.

Subpart H - Materials Handling, Storage, Use, and Disposal

This subpart deals with the proper handling of materials. Stored materials must be stacked, racked, and secured to prevent sliding or falling. All materials must be stored away from aisles and passageways so employees can walk through these areas safely.

Rigging equipment, used for moving materials, is another factor associated with material handling. Every time material is moved, the equipment should be inspected prior to moving. This subpart breaks down the weight of materials that can be moved using a wire rope, which must also be inspected before use. There are several types of ropes that can be used for moving materials

Proper disposal is another issue discussed in subpart H. When disposing of materials like scrap wood, such materials should be removed from the work area immediately to prevent employee injuries. Other materials such as oily rags and flammable liquids should be kept in covered containers until they can be removed from the worksite.

Subpart I - Hand and Power Tools

Subpart I of construction standard 29 CFR 1926, concentrates on the proper use and condition of both hand and power tools. Tool regulations on the jobsite apply to both employee-owned and employer-provided tools. This is important to note because so many times contractors and subcontractors are hired to work on the same jobsite.

Subpart I requires all power-operated tools that are designed with guards be equipped with those guards. Additionally, external protrusions such as wheels, pulleys, and shafts should be guarded according to the requirements presented by the American National Standards Institute (ANSI). Employees working near hazards or under hazardous conditions such as falling objects, dust, or harmful fumes will wear the appropriate personal protective equipment at all times.

Special hand tools shall be provided for the employee to prevent body parts from entering a danger zone of any machine. These tools should not be provided in lieu of using the appropriate guards. Large tools designed to operate in a fixed position must be securely anchored to the ground so that there is no potential for walking or tipping. According to Subpart I, hand tools should not be used if they are in any way damaged or unsafe. For example, chisels with mushroomed heads and wrenches with sprung jaws are unacceptable.

Proper use and handling of fuel-powered, pneumatic, and powder-actuated tools is also covered in Subpart I. Only employees who have been trained should operate powder-actuated tools. In addition, both power and hand tools should be maintained and serviced on a regular basis. Jacks are the last item covered under Subpart I. Proper blocking and securing of the lifted part is covered along with maintenance requirements. The standards also require the tagging of faulty or broken jacks so they will not be used by mistake.

Subpart J - Welding and Cutting

Subpart J of 29 CFR 1926 covers the methods and precautions associated with welding and cutting, arc welding, preventing fires, proper ventilation, and welding of materials. Subpart J covers the standards for most methods of welding and cutting and provides the guidelines for minimum worker safety when carrying out those tasks.

This subpart begins with gas welding and cutting. It addresses the proper storage and joist transportation of gas cylinders. In addition, Subpart J notes that damaged or deteriorated cylinders should not be used. Gas cylinders being used should be marked clearly with their names, using letters of at least one inch in height. Hoses used must be clearly distinguishable, and regulators and gauges should be in good working condition. All components should be inspected before being used.

Next, arc welding and cutting is covered. Proper grounding along with suitable cables and electrodes must be used when arc welding or cutting. Operators must be properly instructed on how to safely use the welding equipment and wear the appropriate personal protective equipment.

Fire protection and prevention is covered in the standards regulating welding and cutting. No welding is to be performed near flammable vapors and fumes, paints, or heavy dust. Fire extinguishing equipment must be available and in good working order at all times.

Finally, the subpart addresses acceptable methods of ventilation. The requirements for mechanical ventilation systems are listed in this subpart and it includes notes on confined spaces and toxic substance welding.

Subpart K - Electrical

Subpart K deals with the means of providing electrical power, both permanent and temporary, to a jobsite. Subpart K also states that the electrical standards do not apply to existing permanent installations that were on the site before construction began. There are four major divisions of this subpart including installation safety requirements, safety-related work practices, maintenance and environmental considerations, and requirements for special equipment.

Installation Safety Requirements. Regulations require that all electrical parts be inspected for quality, durability, and appropriateness. Systems must be properly installed and grounded before use. Supports, holders, and equipment must also be grounded with only a few exceptions.

Safety-Related Work Practices. Subpart K describes acceptable and non-acceptable wiring methods for both temporary and permanent electrical installations. Temporary installations must be removed immediately when the work is finished. When applicable, the standards from the *National Electrical Code* (NEC), should be adhered to when performing any electrical wiring.

Maintenance and Environmental Conditions. Receptacles must be designed for their locations. For example, when one installs a receptacle in a damp or wet location, special grounding methods must be used to prevent accidental contact with water. Proper equipment should be used for those receptacles with special voltages such as air conditioners and clothesdryers.

Special Equipment. Power supplies for equipment such as elevators, escalators and moving walks are addressed under Subpart K. There are also requirements for electrical installations in hazardous environments such as where dust, fumes, and flammable vapor exist. Only equipment designed for the specific location should be used. Batteries and battery charging are regulated under this subpart. Proper handling of battery acids and charging equipment is addressed.

Subpart L - Scaffolding

Subpart L begins by noting that scaffolding will not be used unless it is constructed according to regulations. Footings and anchors must be sound and capable of carrying the intended load. No make-shift footings like bricks and concrete blocks can be used. Scaffolding should always be erected under the supervision of a competent person, which is defined in the subpart. Guardrail and toeboard regulations apply to all the open sides of scaffolding. Different types of scaffolding such as wood pole, tube, and outrigger scaffolds; proper construction and bracing; and anchoring requirements for scaffolding are covered in Subpart L.

After erection, all scaffolding must be capable of supporting its intended loads. This is also applicable to mobile scaffolding--scaffolding on wheels-- which is addressed using 1910 standards. Employees should not work on scaffolding when high winds or stormy conditions pose a hazard. Scaffold work areas must be kept clean and free from the accumulation of tools and materials.

Scaffolds that exceed heights of fifty feet may only be erected by or under the supervision of the scaffold's manufacturer or designated agent unless the structure is approved by a licensed professional engineer. Ladders on scaffolding must be properly constructed and have slip-resistant treads. Scaffolding should never be unstable or of shoddy construction.

Subpart M - Floor and Wall Openings

This subpart applies to both temporary and emergency conditions where an employee may be susceptible to the danger of materials falling through roof, floor, or wall openings, or from runways or stairwells. In general, all floor openings will be guarded by a standard railing on all sides except at entrances to the stairway, and by toeboards or a cover. The height or placement of the wall opening shall be guarded by a standard or immediate rail. A standard toeboard or an enclosing screen should be used to protect a bottom-wall opening.

Types of floor holes and openings include skylights, ladderways, and temporary openings, and floor holes should be guarded by standard railings on all exposed sides and by standard toeboards. A ladderway opening should have an entrance that is blocked by a swinging gate or some other mechanism to prevent a person from falling into the entrance. Floor holes may use a cover in place of the standard railings. However, if the cover is not in use, the floor hole must be guarded by these standard railings.

Pits, trap-door openings, and manholes will be guarded by standard covers of normal construction strength. When the cover is not in place for pits or trap doors, the opening must be protected on all exposed sides by removable standard railings. If the manhole cover is not in place, then the opening will be protected by standard railings. Hatchways and chute floors must have standard railings with only one exposed side or removable railings with toe boards on two sides and fixed railings on the other side.

Standard railings consists of a top rail, intermediate rail, posts, and toeboards. The standard for toeboards includes a four-inch minimum height from the top edge to the length of the floor, ramp, platform, or runway. Handrails shall have the same construction as a standard railing with the exception that the handrail is mounted onto some surface and there is no immediate rail involved. Floor openings such as manholes or trenches shall be capable of supporting the maximum intended load. Wall openings must withstand at least two-hundred pounds applied from any direction.

Subpart N - Cranes, Derricks, Hoists, Elevators, and Conveyors

Subpart N deals specifically with the operation of cranes, derricks, hoists, elevators, and conveyors in the construction industry. In this subpart, employees shall comply with the applicable manufacturer's standards for cranes and derricks. However, when there are no manufacturer's standards, the limits are assigned based on the determination of a qualified engineer. Warnings, special hazards, speed limits, load capacities, and additional instructions shall be posted on all equipment. Annual inspections shall be perform on the equipment by a competent person, government official, or by the U.S. Department of Labor.

This standard applies to such items as crawler, locomotive, truck, overhead, gantry, and tower cranes; helicopter operations; material and personnel hoists, elevators; and conveyors. Loads on the overhead cranes must be labeled and legible from the ground floor. The proper clearance must be maintained between moving and rotating structures on the crane and fixed objects. Employees on these cranes shall be protected from falling by guardrails or safety belts.

Mobile cranes may be mounted on barges but shall not exceed the original capacity set by the manufacturer. The use of a crane or derrick to hoist employees on a personnel platform is prohibited, except when the use or dismantling of a worksite would provide more danger because of the worksite's structural design. Load lines shall be capable of supporting seven times the maximum-intended load. Brakes on these cranes should only be used when the personnel platform is in a stationary position. Personnel platforms will be equipped with guardrails, toeboards, access gates, and headroom for employees to stand upright on the platform.

A trial lift must be performed with an unoccupied personnel platform at least once before the lift of the platform is taken from the ground floor. This trial run must be performed immediately before placing personnel on the platform and must be repeated whenever the cranes or derricks are moved or adjusted. A visual inspection of the crane, derricks, rigging, and personnel platform must be performed by a competent person immediately following the trial lift to determine if any defects or adverse effects have occurred.

Helicopter cranes must comply with the regulations set by the Federal Aviation Administration. The operator of the helicopter is responsible for determining if the load's weight, size, and manner that it is attached to the helicopter can be done safely. Static charge must be eliminated with a grounding device before personnel touch the load being suspended by the helicopter. Visibility is essential for the helicopter operator so he can keep in contact with the ground crew who is sending signals to the operator. When approaching or leaving a helicopter, all personnel must remain in full view of the pilot. Constant communication is essential between the pilot, the ground crew, and other employees to prevent accidents.

All hoists must comply with the manufacturer's specifications, but when they are not available, the limitations of the equipment shall be based on the determination of a professional engineer. Operating procedures should include a signal system, allowable line speeds, and a sign labeling "No Riders Allowed." Hoist cars shall be used in lifting personnel and will be permanently enclosed on all sides and at the top. Safety mechanisms should be capable of stopping or starting the car at any time. Before placing them into service, hoists must be tested, inspected, and constantly maintained on a weekly basis. Employees using conveyors will sound an alarm before they start operation and must be capable of stopping it at any time from the operator's station. In a place where employees come into contact with the conveyor, those individuals should be specially protected.

Subpart O - Motor Vehicles and Mechanized Equipment

All equipment not being used at night or near a highway or construction site must have either the appropriate lights, reflectors, or barricades to identify the location of the equipment. All controls on machinery must be in the neutral position unless work is being done with the machines. The neutral position involves the motor being off and the brakes being set, as if the machine were in a parked position. Safety glass will be used in all cabs and protection will be used when removing tires.

Motor vehicles include those vehicles that do not operate in public traffic, but within an off-highway jobsite. These vehicles will have a service, emergency, and parking brake system and shall be in operable condition. The vehicles will have two headlights and taillights, brake lights, a reverse-signal alarm system, and an audible warning system. Those vehicles with cabs will include windshields, power wipers, and a defogging system. Vehicles used to transport employees will have seats firmly secured in the cab with seat belts that meet Department of Transportation Standards.

Industrial trucks shall meet the following requirements: The capacity will be posted on the vehicle so it is clearly visible to the operator. No modifications should be made to the equipment without the consent of the manufacturer. Protection such as overhead guards or safety platforms should be secured to the lifting mechanism. Finally, all industrial trucks must meet all the ANSI standards for design, construction, stability, inspection, testing, maintenance, and operation.

All boilers and piping systems, either part of or on pile-driving equipment, will have overhead protection, guards, and stop blocks. A blocking device must support the weight of the hammer and will be provided for placement in the leads under the hammer while employees work. All employees shall follow the signals given by the designated signalmen and will be kept clear when piling is being hoisted into the leads. When piles are being placed into a pit, the pit shall be angled.

Marine operations include all loading, unloading, moving, handling, etc., on or out of any vessel from a fixed structure onto a shore vessel. Ramps for the access vehicles between barges must be of adequate strength, protected by side boards, and properly secured. Unless employees can step safely across each barge, a safe walkway such as a "Jacob's Ladder" must be provided. This ladder will be of a flat-tread type, hang without slacking, and be maintained and properly secured.

Subpart P - Excavations
This subpart applies to all those open excavations, cuts, cavities, or trenches made in the earth's surface. Details concerning employee exiting and hazardous atmospheres are of primary importance. In addition, instructions are given regarding underground utilities, necessary protective systems, and soil classifications for trenching, cutting, and excavating.

All underground utilities such as water, sewer, and gas must be identified prior to excavating. Any utility companies shall be contacted in order to advise the excavation crew on the location of these utilities and the proper means for encountering the utility lines. When the excavation is made, the utilities shall be protected or removed in order to safeguard employees while they are working. Structural ramps should be designed by a competent person and in accordance with the design of access and egress from the excavations. A means of egress must be located in trench excavations that are four feet deep, and no more than twenty-five feet deep in order for employees to exit and enter the excavation.

Employees must monitor for hazardous atmospheres to prevent the possible exposure to harmful levels and atmospheric contaminants. If oxygen levels are less than 19.5 percent, the atmosphere must be tested before employees can enter. Oxygen- deficiency exposure precautions should include proper ventilation, respiratory protection, and mandatory testing. If hazards arise while employees are present in the excavation, emergency rescue equipment such as a breathing apparatus, safety harnesses, and stretchers should be readily accessible.

Employees will not work in excavations where water has accumulated, unless measures have been taken to remove the water from the excavation. If excavation operations endanger the structure of buildings, support systems such as shoring, bracing, or underpinning shall be used to ensure the building's stability. Excavation should not be conducted under the base of any foundation because it can put employees in danger. Employees shall also be protected from loose soil or equipment that may fall into the excavation pit. Daily inspections of the excavation, it's adjacent areas, and the protective systems shall be made by a competent person prior to the start of work as needed throughout the shift. Appendices A, C, & D cover excavation wall-shoring procedures.

Subpart Q - Concrete and Masonry Construction

This subpart sets forth the requirements that protect all construction employees from the hazards associated with concrete and masonry construction. In addition, this subpart includes the relevant general industry standards as they apply to this standard. Requirements for bullfloat, formwork, lift slab, precast concrete, reshoring, shoring, vertical slip forms, and jacking operations are also discussed in this subpart.

In relation to structural integrity, all drawings and plans should be available at the worksite. Shoring and reshoring is discussed to ensure that maximum strength is achieved. Any design of the shoring should be done by an engineer or a qualified individual. All reinforcing steel should be adequately supported to make it stable. No framework should be removed until the employer determines the concrete is strong enough to support all loads. Plans, specifications, and tests are used to ensure the concrete has gained sufficient strength.

All precast wall units should be adequately braced to prevent overturning until other permanent structures are in place to support them. Any lifting points attached to the precast lift-up walls should be able to support two times the maximum-expected load. In addition, lifting points or attachments on pieces other than lift-up walls must be able to support four times the maximum-expected load. Hardware should be capable of lifting five times the maximum-expected load to be lifted.

All lift slab operations should be designed by an engineer, and plans should be included with the instructions on proper methods of construction. All jacks should be labeled with their rated capacities, and should be capable of lifting at least two-and-one- half times the load being lifted. Jacks should have a safety device that allows them to stop and support the weight if the load becomes too great to lift or if the jack malfunctions. If any jacking point goes farther than one-half inch from level, jacking operations should stop and the problem fixed. There should be an adequate number of jacks present as specified in this section but the number should never exceed fourteen jacks at one time.

Subpart R - Steel Erection

The requirements for flooring, structural steel assembly, bolting, riveting, fitting-up, and plumbing-up are discussed in this subpart. The standard also addresses the Personal Protective Equipment (PPE) standards for steel erection.

Permanent floors should be installed as progress on the building continues. Information is given on the correct process for laying a temporary floor. These floors must be at least two inches thick or thick enough to support the load. They must also be bolted or otherwise tightly connected to the floor. When any beam or structure is being lifted, the hoisting line should not be loosened until there are at least two bolts at each joint that have been tightened with a wrench. Braces are needed on trusses spanning greater than forty feet before the hoisting line is loosened. Guide lines should be attached to long beams being lifted in order to control them.

All materials, tools, rivets, bolts, etc., should be contained so they do not fall. All pneumatic tools should be disconnected from the power source before any adjustments are made to them. Turnbuckles used for plumbing-up should be properly secured and should not unwind under load. Connection points should also be easily accessible. The use of a safety harness is also stressed in this section.

Subpart S - Underground Construction, Caissons, Cofferdams, and Compressed Air

Subpart S applies to the construction of underground tunnels, shafts, chambers, and passageways. This includes the following conditions:

- Entrances and exits to underground/enclosed areas
- Air quality of underground/enclosed area
- Equipment used in these areas
- Flood controls used in these areas
- Environmental conditions including pressurization and flammability

Items such as check-in/check-out, safety instructions, communication requirements, rescue teams, and hazardous classifications for potentially gassy operations are covered.

There should be safe entrances and exits to tunnels and any unused tunnels or chutes are to be sealed. A check-in/check-out procedure should be used to keep a record of the number of people below ground.

Monitoring should be conducted to determine if there is a possibility of a "gassy" situation below, that could cause an explosion. If the person who monitored for gassy situations determines that a hazardous situation exists, the employer is responsible for taking the proper action. If an individual determines that 20 percent of the lower explosive limit for a gas has been detected, then all personnel should be removed from the area except those necessary to resolve the problem.

All electrical devices should be turned off except for the necessary ventilation pumps. Local gas tests should be performed before and during any welding or other hot work. Exposure to toxic substances should be recorded and the proper fresh air should be supplied to all underground areas.

Open flames are prohibited below ground except for welding or other hot operations. All flammable materials that are permitted underground must be moved carefully and properly. Any leaks or spills should be cleaned up immediately. Proper fire extinguishers should be located below ground. Fire-retardant materials may be necessary around or within tunnels.

Proper bracing should be erected to support underground tunnels. When replacing damaged supports, the new ones should be in place before the damaged ones are removed. Any shaft deeper than five feet that workers enter must be encased by steel, concrete, or timber supports, unless the shaft is cut through rock. After any blasting operation, the shaft must be inspected by a competent person to make certain that it is safe.

When a hoist is in use to lift materials through a shaft, the materials should be secured so they will not fall. In some cases, a person may be needed at the bottom of the shaft to keep people clear while a load is being lowered to the bottom. Personnel should not ride the hoist unless it is specifically for human use or the person is inspecting the shaft. Wire rope used in the hoist must be capable of lifting five times the expected load, or the factor recommended by the rope manufacturer, whichever is greater. A record must be kept of the periodic inspections of lifts. Cages for transporting people must be specifically designed for this purpose and must not exceed certain speeds.

Subpart T - Demolition

A survey should be done by a qualified individual of any structure that is to be demolished. The survey is designed to check the integrity of the framework in order to prevent unplanned collapses. In a building damaged by fire, flood, or explosion, supports should be erected for walls and floors. All gas, sewer, steam, and electrical lines should be cut off prior to demolition. If hazards exist from glass or any other material, then the material should be removed. When material is dropped through holes without a chute, there should be a wall surrounding the landing area that meets all specifications. Any holes not used for dropping debris should be substantially covered, and employee entrances to the building should be protected from falling debris.

Only stairs, passageways, and ladders approved for use in the demolition process are to be used in these buildings. These stairs or ladders are to be periodically inspected. Any stairwell being used in a multilevel building must be properly illuminated and have protective roof coverings where necessary.

A chute with an angle steeper than 45 degrees must be completely covered. The bottom of chutes should be closed off when work is not in progress. Top chute openings, should be protected with guard rails and any space between the edge of the chute and the side of the building must be covered. Chutes must be strong enough to handle the falling debris.

In addition, all holes cut in the floor must not exceed a certain size and should not compromise the structural integrity of the rest of the floor. In steel and masonry constructions, loose masonry must be removed before continuing down to the next level. Walkways or ladders must be available for access to scaffolds or walls. Walls which serve as retaining walls for debris must be capable of retaining the debris. Once floors have been removed, walkways at least eighteen-inches wide must be installed. These walkways must be substantial and planks should overlap at least one foot over solid beams.

Floors must be strong enough to support the equipment being used. Barriers should be erected to keep equipment from falling over the side or through holes. The stored material should not exceed the capacity of the floor. Storage space in which material is to be dumped or dropped is to be concealed.

Except for designated people, no one is to be near the area where mechanical demolition is taking place. Any crane and ball must meet certain requirements. All structural steel beams should be cut before using the demolition ball. Continued monitoring should be conducted throughout the demolition process to check structural integrity of other parts of the building.

Subpart U - Blasting and the Use of Explosives

Only authorized and qualified persons can handle and use explosives. No person handling explosives should be under the influence of alcohol or drugs. Explosives should be accounted for at all times and never be abandoned. Fires should not be fought if there is an imminent danger of contact with explosives. Above-ground blasting should be done between sunrise and sunset. Precautions should be taken to prevent accidental discharge of electric blasting caps from other sources of current. Operators or owners should be notified when blasting will be done near power lines. Loading and firing should be done only by competent individuals. Blasters should be competent, in good physical condition, qualified, and be required to provide evidence of their credentials. They should also be knowledgeable in the use of each type of blasting method. No smoking or flame-producing devices are allowed in the area.

Transportation of explosives shall meet the provisions of the Department of Transportation. The electrical system of trucks should be checked weekly. Auxiliary lights are prohibited. Explosive equipment and supplies should be conveyed singly. They should not be transported on locomotives and the materials should be carried in separate containers. The powder car or conveyance should have a reflectorized sign with the word "Explosives."

Explosives and blasting agents should be stored in approved IRS. Materials should be stored separately from explosives and blasting agents. In addition, smoking is not allowed within 50 feet of explosives. Two modes of exit must be identified if explosives are stored underground.

A warning signal should be given prior to the blast. Flagmen should be stationed on highways, and the blaster fixes the time of blasting. Blasting signals should be posted in a place for employees to study.

Detonators and explosives should not be stored in tunnels, shafts, or caissons. Detonators and explosives are carried separately into the working chambers. The blaster or powderman is responsible for the management aspects.

Subpart V - Power Transmission and Distribution

General requirements relating to power transmission and distribution should identify the proper procedures and actions for initial inspections, tests, and determinations. Gloves or gloves with sleeves should be worn for insulation from the energized parts. Employees should maintain the minimum working distances for various voltage ranges. Lines or equipment should be deenergized before repairing, and guards or barriers should be erected as necessary. Employers should provide training for workers so they all familiar with emergency procedures and first aid.

Rubber equipment should be visually inspected and gloves should have an air test done before use. Protective hats and climbing equipment should be worn. Safety lines should be used in emergency rescue situations. Metal or conductive ladders should not be used near energized lines or other electrical equipment.

Mechanical equipment should be visually inspected daily. Aerial lift trucks should be grounded or barricaded when working near energized lines or they should be insulated for the work being performed. For material handling, materials should be examined before unloading can begin. When hauling poles, the load should be secure and have a red flag attached on the back. Materials should not be stored near energized equipment if it can be stored elsewhere. Tag lines should be used to control loads handled by hoisting equipment. Grounding for protection of employees is also very important. New construction cannot begin unless the equipment is grounded or other means have been implemented to prevent contact with energized lines.

Guarding and ventilating street openings to underground lines should be done. Oxygen levels should be monitored for the presence of fumes or gases. Underground facilities such as gas or telephone lines should be protected from damage when exposed to energized parts. Energized facilities should be identified and the protective equipment needed should be specified. Deenergizing of equipment or barricades should be used for the particular situation. It is important for body belts, safety straps, and lanyards to be used by all personnel. The hardware should be forged or pressed steel and tensile tests should be performed to prove reliability. Testing of safety straps, body belts, and lanyards should be performed regularly.

Subpart W - Rollover Protective Structures and Overhead Protection

Rollover protective structures as well as overhead protection are included in this section's regulations. This includes not only heavy construction equipment, but also farming equipment, mostly tractors, used by the general public. There are rollover protective structures (ROPS) for material handling equipment. If a ROPS is removed, it should be remounted with one of equal or better quality. Labeling should include the manufacturer's name, the model number, and the machine make or model. Machines should meet the requirements of the State of California, the U.S Army Corps of Engineers, or the Bureau of Reclamation of the U.S. Department of the Interior.

There are minimum performance criteria for ROPS on designated scrapers, loaders, dozers, graders, and crawler tractors. There should be material, equipment, and tiedown means adequate for the vehicle frame to absorb the applied energy. A testing procedure, that includes the energy absorbing capabilities of ROPS, the support capability, and the low-temperature impact strength of the material used in the ROPS should be performed.

There are protective-frame (ROPS) test procedures and performance requirements for wheel-type agricultural and industrial tractors used in construction. The protective frame is a structure mounted to the tractor that extends above the operator's seat. It is also used to minimize the chance of injury resulting from normal operations. A field upset test, either a static, or dynamic test, is done to determine ROPS performance. The frame, overhead weather shield, fenders, or other parts may be deformed but should not shatter or leave sharp edges.

Overhead protection for operators of agricultural and industrial tractors is also required under subpart W. The purpose of overhead protection is to minimize the possibility of operator injury from falling objects or from upset of the cover itself. Overhead protection may be constructed of a solid material and should not be installed in a way that causes a hazard. A drop test should be performed following either the static or dynamic test. The same frame should then be subjected to a crush test.

Subpart X - Stairways and Ladders

This subpart concerns stairways and ladders used in construction, alteration, repair, and demolition of workplaces. This section deals with the stairways and ladders used during one of the following events:

- Demolition of buildings and site
- Repair of building and site
- Construction of worksites
- Alteration of buildings and site

It also is applied when ladders and stairways are required to be provided for certain circumstances. A stairway or a ladder is required when there is a break in elevation of nineteen inches or more and no ramp, runway, sloped embankment, or personnel hoist has been provided. Employers are required to provide and install stairwells, ladders, and fall-protection systems before employees can begin work.

Stairs should be installed between 30 degrees and 50 degrees from the horizontal. Variations in riser height and tread depth should not be over one-fourth inch in any stairway system. All stairways must be free of hazardous projections and slip-resistant. Stairways having four or more risers or a rising of more than 30 inches, whichever is less, are required to have at least one handrail and one stair rail system along each unprotected edge. Handrails and the top rails of stairwell systems should be capable of withstanding at least 200 pounds at any point along the top edge. Unprotected sides and edges of stairway landings should be provided with guardrail systems.

Ladders should be capable of supporting specified loads without failure. Ladder rungs, cleats, and steps should be parallel, level, and uniformly spaced. The rungs of individual-rung step ladders should be shaped so that employees' feet can't slide off. They should be coated with a skid-resistant material or treated to minimize slipping. Ladders should not be tied together to provide longer sections unless they are designed for that purpose, and ladders should be surfaced in a way to prevent injury caused by punctures or lacerations.

When the total length of a climb equals or exceeds 24 feet, fixed ladders must have one of the following: a ladder safety device, a self-retracting life line and rest platform or a cage or well, and multiple ladder sections. Cages and wells should conform to various specifications. When using a ladder to gain access to an upper- landing service, the ladder side rails must extend at least 3 feet above the upper-landing service or the ladder should be secured at its top to a rigid support that will not deflect. Ladders should be maintained free of oil, grease, and other slipping hazards and should not be loaded beyond their maximum-designed load. They should be used on stable and level surfaces unless secured to prevent an accident. When an employee uses a ladder, she should always be facing it. The area around the top and bottom of a ladder should be kept clear and ladders should never be moved or extended when occupied. The top or top step of a stepladder should never be used as a step. Ladders should be inspected for defects on a periodic basis and after any accident. Repairs should restore the ladder to its original condition before it is returned to use. Single-rail ladders are prohibited for use.

If necessary, employers should provide a training program on the use of ladders and stairways. The employer should ensure that each employee is trained by a competent individual. Employees should be able to understand the nature of fall hazards, the correct procedures for constructing and maintaining fall-protection systems, the correct use of stairways and ladders, and the maximum-intended carrying capacities of ladders. Retraining should be provided for each employee if necessary.

CHECKLIST - FOR SUBPARTS A-X

Subpart A - General
- Does your company have to comply with the construction standards?
- Does your company need to be granted a variance from the safety and health standards?
- Does your company make special accommodations for inspectors?
- Is your company aware of the Secretary's right of entry procedure?

Subpart B - General Interpretations
- Is your contract one of the 58 statutes listed in the Reorganization Plan No. 14?
- Does your company perform work under federal contracts?
- Does your contract entail the use of subcontractors?
- Does your company delegate responsibility to subcontractors in contracts?

Subpart C - General Safety and Health Provisions
- Does your company have accident-prevention programs in place?
- Does your company promote education for prevention of unsafe working conditions?
- Does your company have an elaborate fire protection and prevention program?
- Does your company keep detailed medical records of employees; if so, how long?
- Does your process incorporate any trade secrets that should not be disclosed to doctors?
- Are all of your exits unlocked and free from impediments?

Subpart D - Occupational Health and Environmental Controls
- Does your company have a written HAZCOM program?
- Does your company offer medical assistance or advice?
- Does your company have a sufficient potable water system?
- Do you have any areas of high noise exposure? How loud?
- Does your facility contain any Asbestos of MDA?
- Does your facility have the proper illumination devices?
- Have you incorporated a Process Safety Management System?
- Do you have the proper Personal Safety Equipment on hand?

Subpart E - PPE and Lifesaving Equipment
- Do the employees at your facility use PPE or lifesaving equipment?
- Are your employees trained on the proper use of PPE?

- Does the facility have the protective equipment for its employees?
- Does the facility have the proper safety equipment needed?
- Does the facility provide safety classes for its employees?
- Does the facility provide the proper respiratory protective equipment?

Subpart F - Fire Protection and Prevention
- Does your company have a fire protection and prevention program in place on- site?
- Does the facility have the proper fire extinguisher in areas needed?
- Does the facility provide training to the employees on extinguishers?
- Does the facility have a sufficient water supply to put out a fire?
- Does the facility have storage area for flammable liquids?
- Does the facility detection system meet OSHA standards?

Subpart G - Signs, Signals, and Barricades
- Does the facility have safety signs in place?
- Does the facility have the proper coloring on the signs?
- Do the employees have knowledge of the signs?
- Does the facility have the signs in the proper areas?

Subpart H - Materials Handling, Storage, Use, and Disposal
- Do the employees of your facility use proper techniques and equipment when handling materials on-site?
- Does the facility have materials stored in proper areas not blocking aisles?
- Does the facility have proper ropes and wires when moving materials?
- Does the facility dispose the materials correctly?

Subpart I - Hand and Power Tools
- Is there on a permanent basis, any power tools or hand tools for use by the company on-site?
- Is there machinery that must be in a fixed position requiring anchors?
- Are there fans with blades less than seven feet above the floor?
- Is there equipment requiring a special hand tool or tools to keep fingers, hands, or any other body part from being harmed?

Subpart J - Welding and Cutting
- Do you employ any full- or part-time welders to weld or cut materials on the site?
- Do you store or use gases commonly used in welding operations?
- Is there any welding required on the job for more than just minor repairs?
- Do you have repairs requiring major welding or cutting?

Subpart K - Electrical

- Does the establishment accommodate elevators, escalators, or powered walks?
- Is there any construction on the site using power tools or any temporary power installation?
- Are there circuit breakers on the premises?
- Does the environment consist of hazardous surroundings such as extremely wet, dusty, or vapor-rich conditions?

Subpart L - Scaffolding

- Is there any construction on the grounds requiring scaffolding?
- Is mobile scaffolding used in the operation?
- Are you at any times required to construct any sort of scaffolding to perform certain tasks?
- Are there any renovations or remodeling that would require scaffolding to be erected on the premises?

Subpart M - Floor and Wall Openings

- Does the facility have floor openings or floor holes?
- If there are wall openings in a facility, do they consist of more than a drop of four feet?
- Have you checked the guidelines pertaining to the width of your roof in Appendix A of Subpart M of the Construction Standards in 29 CFR 1926?

Subpart N - Cranes, Derricks, Hoists, Elevators, and Conveyors

- Are cranes of any type being used in your facility?
- Does your facility use hoists for materials, personnel, or overhead objects that are hard to reach?
- Are there conveyor systems located anywhere in your facility and, if so, do they send a warning signal before starting their operation?

Subpart O - Motor Vehicles, Mechanized Equipment

- Are there any motorized vehicles being used at your facility?
- Is any material-handling, earth-moving, or pile-driving equipment used at your company?
- Are operations occurring between moored vessels and your company's personnel or equipment?
- If motorized vehicles are present on-site, are they equipped with seat belts and adequate safety protection?
- Is your facility near a body of water, and, if so, are there ongoing operations between the land vessel and the marine vessel?

Subpart P - Excavations

- Will there be any forms of excavations at your facility?
- If excavations are present at your place of business, have you considered the location of underground utilities, means of access and egress, and the possibility of hazardous atmospheres prior to excavating on-site?
- What type of support and protective systems are going to be utilized when dealing with the excavations, safety mechanisms, and protection from cave-ins?
- Has the soil at the desired location for the excavation been classified?

Subpart Q - Concrete and Masonry Construction

- Does your company ever perform work such as lift slab, precast concrete, shoring, jacking, or similar operations?

Subpart R - Steel Erection

- Does your company erect steel buildings or perform other steel erection operations?

Subpart S - Underground Construction, Caisson, Cofferdams, and Compressed Air

- Does your company construct underground tunnels, shafts, chambers, or passageways?

Subpart T - Demolition

- Does your company demolish existing structures via mechanical or explosive means?

Subpart U - Blasting and the Use of Explosives

- Does your company use explosives for blasting or other purposes?
- Is the blaster qualified by training, knowledge, or experience in the field of explosives?
- In transporting explosives, does the vehicle have placards on all four sides?
- Are explosives and blasting agents being stored separately?
- Are safety fuses being used in areas where extraneous electricity is present?
- When blasting underwater, are vessels at least 1,500 feet away?

Subpart V - Power Transmission and Distribution

- Does your company erect or improve electrical transmission lines and equipment?

- Are gloves being worn for insulation and are employees maintaining minimum working distances for different voltages?
- Has the equipment been grounded or has other means been implemented to prevent contact with energized lines?
- Have the oxygen levels and the presence of fumes or gases been tested before working on underground lines?
- Are the equipment and rigging being inspected regularly?

Subpart W - Rollover Protective Structures and Overhead Protection

- Is your company's heavy equipment in good condition and equipped with roll cages?
- Is the equipment certified?
- Are the material, equipment, and tiedown means adequate for the vehicle to absorb the applied energy?
- Does the labeling include the manufacturer's name, the model number, and the machine make or model?
- Have a drop test and a crush test been done?

Subpart X - Stairways and Ladders

- Are stairways and ladders used within your facility?
- Are stairways and ladders not being used in your facility when they should be in use?
- Is there a break in elevation of 19 inches or is there no ramp, runway, sloped embankment, or personnel hoist provided?
- Has the employer installed a fall-protection system?
- Is the stairway or ladder free of hazardous objects and treated to minimize slipping?
- Are the unprotected sides and edges of stairways provided with guardrail systems?
- Are ladders being used on stable, level surfaces or have they been secured?

Appendix C

Anthropometric Data

Table of Contents

Table C1 - Standing Body Dimensions (From MIL-STD-1472D)

Percentile Values in Centimeters

	5th Percentile			95th Percentile		
	Ground Troops	Aviators	Women	Ground Troops	Aviators	Women
Weight (kg)	55.5	60.4	46.4	91.6	96.0	74.5
Standing Body Dimensions						
1. Stature	162.8	164.2	46.4	185.6	187.7	174.1
2. Eye Height (standing)	151.1	152.1	152.4	173.3	175.2	162.2
3. Shoulder (acromiale) Height	133.6	133.3	140.9	154.2	154.8	143.7
4. Chest (nipple) Height	117.9	120.8	123.0	136.5	138.5	127.8
5. Elbow (radiale) Height	101.0	104.8	94.9	117.8	120.0	110.7
6. Fingertip (dactylion) Height		61.5			73.2	
7. Waist Height	96.6	97.6	93.1	115.2	115.1	110.3
8. Crotch Height	76.3	74.7	68.1	91.8	92.0	83.9
9. Gluteal Furrow Height	73.3	74.6	66.4	87.7	88.1	81.0
10. Kneecap Height	47.5	46.8	43.8	58.6	57.8	52.5
11. Calf Height	31.1	30.9	29.0	40.6	39.3	36.6
12. Functional Reach	72.6	73.1	64.0	90.9	87.0	80.4
13. Functional Reach, Extended	84.2	82.3	73.5	101.2	97.3	92.7

Percentile Values in Inches

	Ground Troops	Aviators	Women	Ground Troops	Aviators	Women
Weight (kg)	122.4	133.1	102.3	201.9	211.6	164.3
Standing Body Dimensions						
1. Stature	64.1	64.6	60.0	73.1	73.9	68.5
2. Eye Height (standing)	59.5	59.9	55.5	68.2	69.0	63.9
3. Shoulder (acromiale) Height	52.6	52.5	48.4	60.7	60.9	56.6
4. Chest (nipple) Height	46.4	47.5	43.0	53.7	54.5	50.3
5. Elbow (radiale) Height	39.8	41.3	37.4	46.4	47.2	43.6
6. Fingertip (dactylion) Height		24.2			28.8	
7. Waist Height	38.0	38.4	36.6	45.3	45.3	43.4
8. Crotch Height	30.0	29.4	26.8	36.1	36.2	33.0
9. Gluteal Furrow Height	28.8	29.4	26.2	34.5	34.7	31.9
10. Kneecap Height	18.7	18.4	17.2	23.1	22.8	20.7
11. Calf Height	12.2	12.2	11.4	16.0	15.5	14.4
12. Functional Reach	28.6	28.8	25.2	35.8	34.3	31.7
13. Functional Reach, Extended	33.2	32.4	28.9	39.8	38.3	36.5

Table C2 - Seated Body Dimensions (From MIL-STD-1472D)

Percentile Values in Centimeters

	5th Percentile			95th Percentile		
	Ground Troops	Aviators	Women	Ground Troops	Aviators	Women
Seated Body Dimensions						
14. Vertical Arm Reach, Sitting	128.6	134.0	117.4	147.8	153.2	139.4
15. Sitting Height, Erect	83.5	85.7	79.0	96.9	98.6	90.9
16. Sitting Height, Relaxed	81.5	83.6	77.5	94.8	96.5	89.7
17. Eye Height, Sitting Erect	72.0	73.6	67.7	84.6	86.1	79.1
18. Eye Height, Sitting Relaxed	70.0	71.6	66.2	82.5	84.0	77.9
19. Mid-shoulder Height	56.6	58.3	53.7	67.7	69.2	62.5
20. Shoulder Height, Sitting	54.2	54.6	49.9	65.4	65.9	60.3
21. Shoulder-Elbow Length	33.3	33.2	30.8	40.2	39.7	36.6
22. Elbow-Grip Length	31.7	32.6	29.6	38.3	37.9	35.4
23. Elbow-Fingertip Length	43.8	44.7	40.0	52.0	51.7	47.5
24. Elbow Rest Height	17.5	18.7	16.1	28.0	29.5	26.9
25. Thigh Clearance Height		12.4	10.4		18.8	17.5
26. Knee Height, Sitting	49.7	48.9	46.9	60.2	59.9	55.5
27. Popliteal Height	39.7	38.4	38.0	50.0	47.7	45.7
28. Buttock - Knee Length	54.9	55.9	53.1	65.8	65.5	63.2
29. Buttock - Popliteal Length	45.8	44.9	43.4	54.5	54.6	52.6
30. Buttock - Heel Length		46.7			56.4	
31. Functional Leg Length	110.6	103.6	99.6	127.7	120.4	118.6

Percentile Values in Inches

	Ground Troops	Aviators	Women	Ground Troops	Aviators	Women
Seated Body Dimensions						
14. Vertical Arm Reach, Sitting	50.6	52.8	46.2	58.2	60.3	54.9
15. Sitting Height, Erect	32.9	33.7	31.1	38.2	38.8	35.8
16. Sitting Height, Relaxed	32.1	32.9	30.5	37.3	38.0	35.3
17. Eye Height, Sitting Erect	28.3	30.0	26.6	33.3	33.9	31.2
18. Eye Height, Sitting Relaxed	27.6	28.2	26.1	32.5	33.1	30.7
19. Mid-Shoulder Height	22.3	23.0	21.2	26.7	27.3	24.6
20. Shoulder Height, Sitting	21.3	21.5	19.6	25.7	25.9	23.7
21. Shoulder-Elbow Length	13.1	13.1	12.1	15.8	15.6	14.4
22. Elbow-Grip Length	12.5	12.8	11.6	15.1	14.9	14.0
23. Elbow-Fingertip Length	17.3	17.6	15.7	20.5	20.4	18.7
24. Elbow Rest Height	6.9	7.4	6.4	11.0	11.6	10.6
25. Thigh Clearance Height		4.9	4.1		7.4	6.9
26. Knee Height, Sitting	19.6	19.3	18.5	23.7	23.6	21.8
27. Popliteal Height	15.6	15.1	15.0	19.7	18.8	18.0
28. Buttock - Knee Length	21.6	22.0	20.9	25.9	25.8	24.9
29. Buttock - Popliteal Length	17.9	17.7	17.1	21.5	21.5	20.7
30. Buttock - Heel Length		18.4			22.2	
31. Functional Leg Length	43.5	40.9	39.2	50.3	47.4	46.7

Table C3 - Body Depth and Breadth Dimensions (From MIL-STD-1472D)

Percentile Values in Centimeters

	5th Percentile			95th Percentile		
	Ground Troops	Aviators	Women	Ground Troops	Aviators	Women
Depth and Breadth Dimensions						
32. Chest Depth	18.9	20.4	19.6	26.7	27.8	27.2
33. Buttock Depth		20.7	18.4		27.4	24.3
34. Chest Breadth	27.3	29.5	25.1	34.4	38.5	31.4
35. Hip Breadth, Standing	30.2	31.7	31.5	36.7	38.8	39.5
36. Shoulder (bideltoid) Breadth	41.5	43.2	38.2	49.8	52.6	45.8
37. Forearm - Forearm Breadth	39.8	43.2	33.0	53.6	60.7	44.9
38. Hip Breadth, Sitting	30.7	33.3	33.0	38.4	42.4	43.9
39. Knee-to-Knee Breadth		19.1			25.5	

Percent Value In Inches

	Ground Troops	Aviators	Women	Ground Troops	Aviators	Women
Depth and Breadth Dimensions						
32. Chest Depth	7.5	8.0	7.7	10.5	11.0	10.7
33. Buttock Depth		8.2	7.2		10.8	9.6
34. Chest Breadth	10.8	11.6	9.9	13.5	15.1	12.4
35. Hip Breadth, Standing	11.9	12.5	12.4	14.5	15.3	15.6
36. Shoulder (bideltoid) Breadth	16.3	17.0	15.0	19.6	20.7	18.0
37. Forearm - Forearm Breadth	15.7	17.0	13.0	21.1	23.9	17.7
38. Hip Breadth, Sitting	12.1	13.1	13.0	15.1	16.7	17.3
39. Knee-to-Knee Breadth		7.5			10.0	

Table C4 - Body Circumference and Surface Dimensions (From MIL-STD-147

Percentile Values in Centimeters

	5th Percentile			95th Percentile		
	Ground Troops	Aviators	Women	Ground Troops	Aviators	Women
Circumferences						
40. Neck Circumference	34.2	34.6	29.9	41.0	41.6	36.7
41. Chest Circumference	83.8	87.5	78.4	105.9	109.9	100.2
42. Waist Circumference	68.4	73.5	59.5	95.9	101.7	83.5
43. Hip Circumference	85.1	87.1	85.5	106.9	108.4	106.1
44. Hip Circumference, Sitting		97.0	87.7		119.3	110.8
45. Vertical Trunk Circumference, Standing	150.6	156.3	142.2	178.6	181.9	168.3
46. Vertical Trunk Circumference, Sitting		150.4	134.8		175.0	161.0
47. Arm Scye Circumference	39.6	39.9	33.6	50.3	53.0	41.7
48. Biceps Circumference, Flexed	27.0	27.8	23.2	37.0	36.9	30.8
49. Elbow Circumference, Flexed		28.5	23.5		34.2	30.0
50. Forearm Circumference, Flexed	26.1	26.3	22.2	33.1	33.1	27.5
51. Wrist Circumference	15.7	15.3	13.6	18.6	19.2	16.2
52. Upper Thigh Circumference	48.1	49.6	48.7	63.9	66.9	64.5
53. Calf Circumference	31.6	33.3	30.6	41.2	41.3	39.2
54. Ankle Circumference	19.3	20.0	18.7	25.2	24.8	23.3
55. Waist Back Length	39.2	42.4	36.7	50.8	50.9	45.4
56. Waist Front Length	36.1	35.7	30.5	46.2	44.2	41.4

Percentile Values in Inches

	Ground Troops	Aviators	Women	Ground Troops	Aviators	Women
Circumferences						
40. Neck Circumference	13.5	13.6	11.8	16.1	16.4	14.4
41. Chest Circumference	33.0	34.4	30.8	41.7	43.3	39.5
42. Waist Circumference	26.9	28.9	23.4	37.8	40.0	32.9
43. Hip Circumference	33.5	34.3	33.7	42.1	42.7	41.8
44. Hip Circumference, Sitting		38.2	34.5		47.0	43.6
45. Vertical Trunk Circumference, Standing	59.3	61.6	56.0	70.3	71.6	65.5
46. Vertical Trunk Circumference, Sitting		59.2	53.1		68.9	63.4
47. Arm Scye Circumference	15.6	15.7	13.2	19.8	20.9	16.4
48. Biceps Circumference, Flexed	10.6	11.0	9.1	14.6	14.5	12.1
49. Elbow Circumference, Flexed		11.2	9.2		13.5	11.8
50. Forearm Circumference, Flexed	10.3	10.4	8.7	13.0	13.0	10.8
51. Wrist Circumference	6.2	6.0	5.4	7.3	7.6	6.4
52. Upper Thigh Circumference	18.9	19.5	19.2	25.1	26.3	25.4
53. Calf Circumference	12.4	13.1	12.0	16.2	16.3	15.4
54. Ankle Circumference	7.6	7.9	7.4	9.9	9.7	9.2
55. Waist Back Length	15.4	16.7	14.4	20.0	20.0	17.9
56. Waist Front Length	14.2	14.1	12.0	18.2	17.4	16.3

Table C5 - Hand and Foot Dimensions (From MIL-STD-1472)

Percentile Values in Centimeters

	5th Percentile			95th Percentile		
	Ground Troops	Aviators	Women	Ground Troops	Aviators	Women
Hand Dimensions						
57. Hand Length	17.4	17.7	16.1	20.7	20.7	20.0
58. Palm Length	9.6	10.0	9.0	11.7	11.9	10.8
59. Hand Breadth	8.1	8.2	6.9	9.7	9.7	8.5
60. Hand Circumference	19.5	19.6	16.8	23.6	23.1	19.9
61. Hand Thickness		2.4			3.5	
Foot Dimensions						
62. Foot Length	24.5	24.4	22.2	29.0	29.0	26.5
63. Instep Length	17.7	17.5	16.3	21.7	21.4	19.6
64. Foot Breadth	9.0	9.0	8.0	10.9	11.6	9.8
65. Foot Circumference	22.5	22.6	20.8	27.4	27.7	24.5
66. Heel-Ankle Circumference	31.3	30.7	28.5	37.0	36.3	33.3

Percent Values in Inches

	Ground Troops	Aviators	Women	Ground Troops	Aviators	Women
Hand Dimensions						
57. Hand Length	6.85	6.98	6.32	8.13	8.14	7.89
58. Palm Length	3.77	3.92	3.56	4.61	4.69	4.24
59. Hand Breadth	3.20	3.22	2.72	3.83	3.80	3.33
60. Hand Circumference	7.68	7.71	6.62	9.28	9.11	7.82
61. Hand Thickness		0.95			1.37	
Foot Dimensions						
62. Foot Length	9.65	9.62	8.74	11.41	11.42	10.42
63. Instep Length	6.97	6.88	6.41	8.54	8.42	7.70
64. Foot Breadth	3.53	3.54	3.16	4.29	4.58	3.84
65. Foot Circumference	8.86	8.91	8.17	10.79	10.62	9.65
66. Heel-Ankle Circumference	12.32	12.08	11.21	14.57	14.30	13.11

Index

Environmental and Health/Safety References

Total Quality for Safety and Health Professionals
F. David Pierce, a CSP and a CIH, shows you how to apply concepts - proven and successful - to your safety management program to achieve increased productivity, lowered costs, reduced inventories, improved quality, increased profits, and raised employee morale.
Hardcover, 244 pages, June '95, ISBN: 0-86587-462-X **$59**

Pollution Prevention Strategies and Technologies
This book is an indispensible guide to understanding pollution prevention policies and regulatory initiatives designed to reduce wastes. *Hardcover, Index, 484 pages, Oct '95, ISBN: 0-86587-480-8* **$79**

Environmental Audits, 7th Edition
This is the most comprehensive manual available on environmental audits! Completely updated, it provides you with all the step-by-step guidance you need on how to begin - and manage - a successful audit program for your facility. Includes over 50 pages of exercises that cover Audit Verification, Interviewing Skills, Management Assessments, Report Writing, and Audit Conferences.
Softcover, approx. 500 pages, Mar '96, ISBN: 0-86587-525-1 **$79**

Toxicology Handbook
Get a basic, non-technical understanding of toxicological principles and have the relevant EPA Key Guidance and Implementation Documents highlighted throughout.
Softcover, 180 pages, Sept '86, ISBN: 0-86587-142-6 (Code 714) **$65**

Environmental Law Handbook, 13th Edition
Includes changes in major federal environmental laws. Details how those laws affect the regulated community.
Hardcover, Index, 550 pages, Mar '95, ISBN: 0-86587-450-6 **$79**

"So You're the Safety Director!"
An Introduction to Loss Control and Safety Management
This book concentrates on your role in evaluating, managing, and controlling your company's losses and handling the OSHA compliance process.
Softcover, Index, 186 pages, Oct '95, ISBN: 0-86587-481-6 **$45**

Safety Made Easy: A Checklist Approach to OSHA Compliance
This new book provides a simple way of understanding your requirements under the complex maze of the Occupational Safety and Health Administration regulations. The authors have created safety and health checklists for compliance organized alphabetically by topic.
Softcover, 192 pages, June '95, ISBN: 0-86587-463-8 **$45**

Understanding Workers' Compensation: A Guide for Safety and Health Professionals
This book is designed to help you understand how the workers' comp system works, and provides a basic understanding of injury prevention, types of injuries, and cost containment strategies.
Softcover, 250 pages, July '95, ISBN: 0-86587-464-6 **$45**

Government Institutes • 4 Research Place • Rockville, MD 20850 • USA • (301) 921-2355

Environmental and Health/Safety References

Chemical Information Manual, 3rd Edition (Book and Disk Format)
This database contains essential data for over 1,400 chemical substances. The following information is available to you: proper identification synonyms, OSHA exposure limits, description and physical properties; carcinogenic status, health effects and toxicology, sampling and analysis.
Softcover, 400 pages, Aug '95, ISBN: 0-86587-469-7 **$99**
Also available on disk!
3.5" Floppy Disk for Windows, #4070 **$99**

Occupational Safety and Health Administration Technical Manual, 4th Edition

This inspection manual is used nationwide by the U.S. Occupational Safety and Health Administration's inspectors in checking industry compliance with OSHA require-ments. *Softcover, 400 pages, Feb '96, ISBN: 0-86587-511-1* **$85**
U.S. Government Document

Exposure Factors Handbook, Review Draft
The U.S. Environmental Protection Agency uses this document to develop pesticide tolerance levels, assess industrial chemical risks, and to undertake Superfund site assessments and drinking water health assessments.
Softcover, 866 pages, Nov '95, ISBN: 0-86587-509-X **$125**

Ergonomic Problems in the Workplace: A Guide to Effective Management
The valuable insights you'll gain from this new book will help you develop and implement your own successful ergonomics program.
Softcover, 256 pages, Sept '95, ISBN: 0-86587-474-3 **$59**

Product Side of Pollution Prevention: Evaluating the Potential for Safe Substitutes
This report focuses on safe substitutes for products that contain or use toxic chemicals in their manufacturing process.
Softcover, 240 pages, Sept '95, ISBN: 0-86587-479-4 **$69**

Sampling, Analysis & Monitor-ing Methods: A Guide to EPA Requirements
This book provides a guide for determining which chemicals have sampling, analysis, and monitor-ing requirements under U.S. environmental laws and regula-tions, and where those testing and sampling methods can be found.
Softcover, 256 pages, Sept '95, ISBN: 0-86587-477-8 **$65**

Emergency Planning & Manage-ment: *Ensuring Your Company's Survival in the Event of a Disaster*
This book will help you assess your exposure to disasters and prepare emergency response, preparedness, and recovery plans for your facilities, both to comply with OSHA and EPA requirements and to reduce the risk of losses to your company.
Softcover, 192 pages, Nov '95, ISBN: 0-86587-505-7 **$59**

Environmental, Health & Safety CFRs on CD ROM
Now you can scan EPA, OSHA and Hazmat regulations in seconds. Search on a regulation number, word, or phrase, and print or save the results!
Single Issue (most recent quarterly release) #4018 **$395** + **$5** shipping & handling

One-Year Subscription (receive an updated CD quarterly, beginning with the most recent release) #4000 **$998**

Government Institutes • 4 Research Place • Rockville, MD 20850 • USA • (301) 921-2355